Lecture Notes in Statistics

Edited by D. Brillinger, S. Fienberg, J. Gani, J. Hartigan, and K. Krickeberg

23

Diane Griffin Saphire

Estimation of Victimization Prevalence Using Data from the National Crime Survey

Springer-Verlag
New York Berlin Heidelberg Tokyo 1984

Author

Diane Griffin Saphire
Trinity University, Department of Mathematics
715 Stadium Drive, San Antonio, TX 78284, USA

AMS Subject Classification (1980): 62D05

ISBN 3-540-96020-1 Springer-Verlag Berlin Heidelberg New York Tokyo
ISBN 0-387-96020-1 Springer-Verlag New York Heidelberg Berlin Tokyo

Library of Congress Cataloging in Publication Data Saphire, Diane Griffin. Estimation of victimization prevalence using data from the National crime survey. (Lecture notes in statistics; 23) Bibliography: p. 1. Victims of crimes – Forcasting – United States – Mathematical models. 2. Victims of crimes – United States – Longitudinal studies. 3. Victims of crimes surveys – United States – Mathematical models. 4. Criminal statistics – United States – Mathematical models. I. Title. II. Series: Lecture notes in statistics (Springer-Verlag); v. 23.
HV6250.3.U5S22 1984 364.1'0723 84-13909

Printed in Germany

Printing and binding: Beltz Offsetdruck, Hemsbach / Bergstr.
2146/3140-543210

Preface

The National Crime Survey is a sample survey of housing units conducted by the U.S. Bureau of the Census. All eligible occupants of a sampled unit are interviewed every six months (for up to seven interviews) about victimizations that they have experienced during the previous six months. In this monograph several longitudinal analyses are performed using a subsample of the data covering the years 1973 through 1975. In particular, several methods of estimating the proportion of units that are crime-free for a given year, denoted by θ, are discussed.

First, several ad hoc, as opposed to model-based, estimators of θ are discussed, including those used by the Bureau of Justice Statistics. We find models under which these estimators are consistent for θ. One such model fits the data very well.

A superpopulation approach to the estimation of θ is then taken, assuming that the nonresponse and sampling mechanisms are ignorable. Three models are fit to the data: i) a homogeneous Bernoulli model, under which victimization is independent from month to month ii) a correlated Bernoulli model, under which victimization in any two months has positive correlation ρ, and iii) a two-state Markov model with states "victimized" and "crime-free". The correlated Bernoulli model is found to be very inadequate. The other two models fit the 1975 data well, but have rather poor fits to the 1973 and 1974 data. Rotation group biases are conjectured to be the cause of these poor fits.

By separating the units with no nonresponse from those with at least one nonresponse and by estimating θ for each of these groups, units with at least one nonresponse are found to have a lower probability of being crime-free for the year. To incorporate nonresponse in the modelling, a logit model with explanatory variables "months of information" and "number of noninterviews" is fit and three alternative estimators of θ are computed.

In general, the estimates of θ lie in the range .65 to .70 with estimated standard errors of about .02. The estimates tend to increase slightly from 1973 to 1975.

* * *

Much of the work presented here was done while I was at Carnegie-Mellon University and I would like to express my gratitude to the faculty, staff and students there for their help in the preparation of this monograph. Many thanks are due to Stephen Fienberg for his support and guidance and for his statistical and editorial assistance. I would like to thank Bill Eddy for all of his statistical and computing help, and especially for his warm words of encouragement. I am indebted to Luke Tierney for his assistance with the work presented in Chapter 3 and with the Markov model variances, as well as his comments on the rest of my work. Thanks are due to Mark Schervish for prompt and insightful responses during both the research and writing stages. Suggestions from Robert Johnson, Alfred Blumstein, and Charles Alexander were very valuable to me in the writing of the final version.

The research presented here was funded by the Bureau of Justice Statistics and the National Institute of Justice, both of the Office of Justice Assistance Research and Statistics, U.S. Department of Justice, under grants J-LEAA-015-79 and 81-IJ-CX-0087, respectively.

TABLE OF CONTENTS

CHAPTER 1

INTRODUCTION

1.1. Cheery Indicators

During the 1960's and 1970's there was an increasing interest in crime in the United States. Police reports showed that crime was rising and public opinion polls showed that public awareness of the crime problem was growing. In order to study crime from the perspective of the victim, attention was turned to sample surveys on victimization.

During the mid-1960's, the Crime Commission, more formally known as the President's Commission on Law Enforcement and Administration of Justice, sponsored research in the area of victimization surveys. In particular, the Crime Commission funded the National Opinion Research Center (NORC) to conduct a survey designed to produce estimates of the true national victimization rate, as opposed to rates computed using only information reported to the police. Although there were difficulties with the NORC survey, the data collected clearly showed that there were many more victimizations occurring in the nation than were reported to the police. As a result, the Crime Commission recommended that a National Criminal Justice Statistics Center be established to continue the collection of crime data.

In 1968, the Congress established the Law Enforcement Assistance Administration (LEAA). A Statistical Division was formed within the LEAA to encourage the collection of criminal justice data. In 1969, LEAA began a series of field studies to obtain information about the measurement of victimization. For descriptions and the major findings of these surveys, see Lehnen and Skogan (1981). Based on this information the National Crime Survey (NCS) was developed by the LEAA in conjunction with the Bureau of the Census. The Bureau of the Census conducted the survey for the LEAA until 1980 when the statistical component of the LEAA was succeeded by the Bureau of Justice Statistics (BJS). Since that time, the Bureau of the Census has continued to conduct the survey, but for the BJS.

Persons participating in the NCS are interviewed and asked detailed questions about any victimizations they have experienced within the six months prior to the interview. Most of the statistical analyses that have been done on the resulting data to date have focused on incidence rates, for instance, the number of victimizations per capita. Quite recently, interest at BJS has turned to indicators of the prevalence of victimization, in part to address the public's desire for information on the risk of victimization. Although victimization rates, which are expressed by the BJS as crimes per 1000 households or crimes per 1000 persons, indicate the volume of crime in the population, they do not show whether the crime is widely spread among the population or highly concentrated. In March of 1981, the BJS issued its first report (U.S. Department of Justice, 1981a) on the prevalence of crime, in which the key quantity estimated was the percentage of households touched by crime in the given year. A household whose members had been victimized, for instance, three times in the year would be counted as one victimized household in this prevalence estimate rather than contributing three crimes to an incidence rate. A subsequent report (U.S. Department of Justice, 1982) appeared within the past year and compares the percentage of households touched by crime for several different subsets of the population and several types of crime.

It has been suggested by Albert Biderman that the fraction of households that are *not* victimized in a year would be a more pleasant prevalence measure than the fraction victimized, and he referred to this quantity as a *cheery indicator* of crime. In addition, the concept of absolutely no victimizations is more precise than the concept of one or more victimizations of some type. Here we focus on the estimation of the probability that a household is crime-free for a year. Throughout this monograph, we will denote this quantity by the symbol θ.

In the rest of Chapter 1 we discuss the survey, the data, and problems of nonresponse. Chapter 2 describes four intuitive estimators of θ including the BJS estimators. After making several fairly weak assumptions, three of these estimators are shown in Chapter 3 to be consistent only under unreasonable models while one is consistent under a model which seems to fit the data well. In Chapter 4 we take a modelling approach to estimating θ and find that the models discussed there do not seem to accurately describe the 1973 and 1974 data, although they do much better for the 1975 data. Comparing estimates computed from a one percent sample of the data, we find that the cheery indicator is not decreasing as we might have expected but may actually be increasing slightly. Chapter 5 addresses the question of whether or not θ is different for the nonrespondents and the respondents. The analyses in Chapters 2 through 5 are performed with the place of residence being the unit of interest. In Chapter 6 some ideas for performing these analyses with the household as the unit of interest are discussed.

1.2. Aims of the NCS

In July of 1972, after extensive pretesting, the Bureau of the Census began conducting for the Law Enforcement Assistance Administration of the U.S. Department of Justice a survey of respondents in selected housing units throughout the United States to enquire about personal and household victimization. This survey has been conducted on a continuing basis since that time and is known as the National Crime Survey. In addition to this national survey, the NCS at one time included additional city surveys and national commercial surveys but these have been discontinued (see Pennick and Owens, 1976). For the remainder of this work, we discuss only the national sample survey of housing units and we refer to it simply as the NCS.

According to the BJS (U.S. Department of Justice, 1981b), the NCS was designed to achieve three primary objectives: "to develop detailed information about the victims and consequences of crime, to estimate the numbers and types of crimes not reported to police, and to provide uniform measures of selected types of crimes in order to permit reliable comparisons over time and among areas." By interviewing members of families occupying housing units from a representative probability sample, our information about crimes is not limited to those crimes that are reported to the police. Thus we no longer need to rely solely on law enforcement agency records (e.g. as incorporated in the FBI's Uniform Crime Reports(U.S. Department of Justice (1981c)) for crime data. Since roughly 50% of the victimizations reported in NCS interviews have not been reported to the police, the information gathered as part of the NCS is essential to meeting the aims of the BJS. For a more detailed description of the proportion of crimes not reported to the police, see U.S. Department of Justice (1981c).

1.3. The Sample Design and Rotation Scheme

The NCS sample design is, in general, a stratified multistage cluster sample of housing units (HUs). The U.S. has been divided by the Bureau of the Census into 1,931 Primary Sampling Units (PSUs) formed from counties or groups of contiguous counties. These are identical to those for the Current Population Survey. The PSUs are grouped into 376 strata. Of these, 156 strata consist of only one PSU each and these are referred to as self-representing PSUs. The remaining 220 strata are formed by combining PSUs with similar characteristics such as geographic region, population density, proportion non-white, etc. The strata are formed so that their 1970 population sizes are approximately equal. From each of these 220 non-self-representing strata, one PSU is selected with probability proportional to the population size. The resulting 220 non-self-representing PSUs together with the 156 self-representing PSUs complete the first stage of sampling. This first stage is currently undergoing substantial revision on the basis of the 1980 decennial census, and a new design will be implemented beginning some time in 1984.

The second stage of sampling consists of selecting enumeration districts from each chosen PSU. Enumeration districts (EDs) are geographic areas used in the 1970 Census that generally have well defined boundaries and contain on the average about 300 households. The EDs are selected with probabilities proportional to their 1970 population sizes.

The third and final stage of sampling involves subdividing each chosen ED into clusters of about four housing units (HUs) and selecting a sample of these clusters.

In addition to HUs chosen in this manner, a sample of units that were constructed after 1970 and units in group quarters such as boarding houses and dormitories is selected and added to the main sample. The basic sample is divided into six subsamples or rotation groups of about 9,000 HUs each. The rotation groups are numbered from 1 through 6 within each sample. Every six months a new rotation group enters the sample and the "oldest" existing rotation group from the previous sample is dropped. Each rotation group is divided into six panels with panel 1 being interviewed in January and July, panel 2 in February and August, etc. This process spreads the workload of the field staff. Each HU is in the survey for three full years for a total of seven interviews. (However, some of the HUs that were rotated into the sample in 1972 and 1973 actually had as many as eight or nine interviews. If the survey would had started by interviewing one panel-rotation group in the first month and rotating one panel-rotation group into the survey at each subsequent month, the number of units being interviewed in the early months would have been very small. To avoid this problem, some units were rotated into the sample before they normally should have.) The data collected at the first of the seven interviews have been used by the BJS and the Bureau of the Census only for bounding purposes, i.e., to establish a time frame intended to avoid duplication of victimization information in subsequent interviews. There is a tendency for respondents to report victimizations as having occurred more recently than they actually did. This phenomenon is referred to as *telescoping*. Dodge (1981) finds that, in a survey of persons who had reported crimes to the police, 19 of 100 later recalled the crime as having occurred more recently than it was recorded as having occurred in the police records. Bounding interviews are used to reduce the number of victimizations that are telescoped into the frame of reference. Thus the data from initial bounding interviews are not incorporated into the official BJS reported rates (either incidence or prevalence) but, as we will see in the following section, many interviews are actually unbounded. In this monograph, we do not attempt to control for telescoping since bounding information is not available for all of the interviews. The information from the bounding interviews that do exist have been incorporated in rates reported here.

Table 1 shows the rotation scheme. For instance, in September, panel 3 in each of the rotation groups 2 through 6 of sample A and 1 and 2 of sample B will be interviewed. In the following March, panel 3 of rotation group 2 in sample A is replaced by panel 3 of rotation group 3 in sample B.

TABLE 1

Rotation Scheme
(The numerical entries represent panel numbers within samples)

	Rotation Group											
	Sample A						Sample B					
Month	1	2	3	4	5	6	1	2	3	4	5	6
Jan	1	1	1	1	1	1	1					
Feb	2	2	2	2	2	2	2					
Mar	3	3	3	3	3	3	3					
Apr	4	4	4	4	4	4	4					
May	5	5	5	5	5	5	5					
June	6	6	6	6	6	6	6					
July		1	1	1	1	1	1	1				
Aug		2	2	2	2	2	2	2				
Sept		3	3	3	3	3	3	3				
Oct		4	4	4	4	4	4	4				
Nov		5	5	5	5	5	5	5				
Dec		6	6	6	6	6	6	6				
Jan			1	1	1	1	1	1	1			
Feb			2	2	2	2	2	2	2			
Mar			3	3	3	3	3	3	3			
Apr			4	4	4	4	4	4	4			
					⋮							

1.4. The Interview

At each interview, a member of the household, usually the head of the household or the spouse of the head of the household, is designated as the household respondent. The questionnaire begins with questions directed to the household respondent about the household, personal characteristics of the respondent, and screening questions about household victimizations (burglary, larceny, and auto theft). These victimizations are crimes that have not been committed against any individual household member in particular but rather against the household in general (for example, an attempted break-in). The screening questions are designed to determine if any household victimizations have occurred during the 6 calendar months preceding the interview. For instance, a household respondent interviewed in July is asked screening questions about the time period of January 1st through June 30th. If he/she responds positively to any one of these questions, the interviewer then asks further questions about the incident and the offender. When the household victimizations, if there are any, are reported, the interviewer asks the respondent a series of screening questions to determine if he/she has experienced a personal victimization during the 6 months prior to the interview. Again, if the respondent answers "yes" to any of these questions, the interviewer asks questions about the incident and the offender. An excerpt of the questionnaire is included here as Appendix A.

When the household respondent's interview has been completed, each remaining household member, age 14 or older, is interviewed individually. The interviewer collects background information about the household member and then asks the series of screening questions, following up with questions about any incidents that are reported. Information about household members aged 12 or 13, members who are physically or mentally unable to answer the questionnaire, or members who are temporarily absent and are not expected to return during the interview month is obtained by a proxy. The proxy may be the household respondent or other knowledgeable household member.

The initial contact with the household is a personal visit, at which time the interviewer obtains interviews from as many of the household members as possible. Subsequent to this initial visit, the interviewers may make telephone call backs to obtain interviews with the remaining eligible household members. In order to save time and money, many of the subsequent interviews may be conducted entirely by telephone. Lehnen and Reiss (1978) note that fewer incidents are reported during telephone interviews than during personal interviews.

The term "household", denoted by HH, is used to refer to the people living in the housing unit and should be distinguished from the term HU which refers to the actual place of residence, i.e. a physical location. If the HH living in the HU at the time of the interview is not the same HH that was living there at the time of the previous interview, no attempt is made by the Bureau of the Census interviewers to locate the original HH. Rather, the HH living in the HU is interviewed and the interviewer notes on the questionnaire that the HH has changed. Thus one HU in the survey may actually, over several interviews, consist of several different HHs. For this reason, using the first interview at a particular HU as a bounding interview, as mentioned in the previous section, may still leave many interviews unbounded since there is no bounding interview for a new HH that moves into the HU during the time that the HU is in the sample.

A problem that may arise during an interview is that in some cases a victim cannot distinguish in time and place among a series of similar victimizations that he/she has experienced. An often cited example of this situation is a woman who is repeatedly beaten by her husband. If she is unable to answer questions about individual beatings, the incidents are recorded as one *series incident* along with the month during which the first incident occurred, the season(s) during which these victimizations occurred and the estimated number of victimizations that are included in the series incident. Victimization rates (e.g. number of victimizations per 1000 HUs) may be greatly underestimated by the total omission of such incidents and, clearly, counting each series incident as one victimization will still lead to an underestimate. For a discussion of the effect of the omission of series victimizations on rates published by the U.S. Department of Justice, see Fienberg (1980). By estimating victimization prevalence rates rather than incidence rates, we eliminate some of the problems associated with these series victimizations. When focussing on prevalence rates, we are only interested in knowing whether or not an HU has been victimized in a given year and the actual number of victimizations is not important to us. Thus the only time a series incident would present any difficulty is if the series were reported as starting near the end of one year but actually extended into the subsequent year. Unfortunately, prior to 1979 we are not definitely able to determine from the recorded questionnaire response whether such a series had occurred. Because of changes made to the survey questionnaire beginning in 1979, this information should be available for recent years; however, as we describe in the next section, the only longitudinally organized NCS data tapes available to us cover the years 1972 through 1975. Eddy et al. (1981b) have computed some of the prevalence estimates presented in later chapters both including and excluding series incidents and have found little (about 1%) difference in the estimates. Because of these findings, in the analyses presented here a series incident has been counted as one incident and no separate analyses excluding series incidents have been performed. (We note that when estimating incident rates, the effect of excluding series incidents is much more severe since as many as 20% of all personal victimizations may be excluded.)

1.5. Data Tapes

The NCS victimization data are publicly available through the Inter-University Consortium for Political and Social Research (ICPSR) at the University of Michigan. These data to date have been stored primarily in cross-sectional form. Each year is divided into four quarters: Jan-Feb-Mar, ..., Oct-Nov-Dec. All information collected at interviews taking place in any given quarter is stored together on one tape. Since each HU is interviewed only once every six months, at most one interview for any given HU appears on each quarterly collection tape. The collection scheme is illustrated in Table 2. From this table, we see that, in order to obtain all the information for a calendar year for any HU, it is necessary to search through 2 quarterly collection tapes if the HU is in panel 1, or 3 tapes if the HU is in any other panel. For example, retrieving the yearly data for an HU in panel 2 involves searching the Jan-Feb-Mar collection tape for that HU's January information, the July-Aug-Sept collection tape for the February through July information, and the subsequent Jan-Feb-Mar collection tape for the August through December information. Since reorganizing NCS collection quarter data in this fashion would be a very time-consuming and cumbersome task, the large majority of the analyses that have so far been performed on the NCS data have been cross-sectional in nature.

There are three types of information collected by the NCS at each interview and stored on the data tapes: household items, individual items for each resident 12 years of age or older, and a crime incident report for each victimization reported at the interview. Because HHs have varying numbers of individuals, and individuals report varying numbers of crime incidents, it is not sensible to store these data in a rectangular array with each row representing an HU, HH, or individual. The ICPSR stores the data in an OSIRIS IV hierarchical file. For a more detailed description of this file see Eddy et al. (1981), Eddy et al. (1982), and Bixby et al. (1979).

TABLE 2

Data Storage

The Xs indicate months of interview.
The outlines indicate data
stored on one quarterly collection tape.

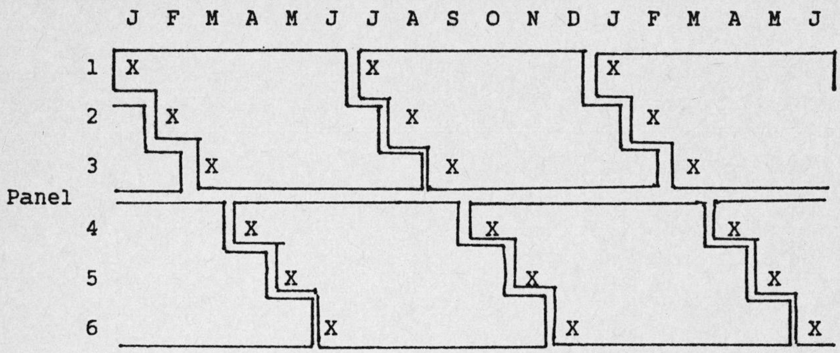

Reiss (1980) has constructed longitudinal files from the quarterly collection tapes for the years 1972-1975. These files make longitudinal analyses feasible and in particular make gathering the data to estimate θ, the proportion of HUs that are not victimized in a year, possible in a straightforward way. These longitudinal data consist of three separate files: one for the HH information, one for the individual information, and one for the crime information. All of the HH information for each of the interviews that an HH has had is stored together on the HH file, and similarly for the individual and victimization information. If information from more than one of these files is needed for an analysis, the HH record can be matched to the records of the persons living in the HH and to any victimizations they reported by the 12 digit identification number which uniquely identifies an HH and appears on the HH, individual and crime records for that HH. This identification number consists of the 2 digit sample number, followed by the 3 digit PSU number, the 1 digit panel number, the 1 digit rotation group number, the 2 digit cluster number (clusters are numbered sequentially within each panel and rotation combination), a 2 digit serial number (which is assigned to HUs within clusters), and the 1 digit HH number (which is assigned sequentially to HHs within the same HU). The first 11 of the 12 digits of the identification number uniquely identify the HU and thus data for HUs, as opposed to HHs, can be gathered from the longitudinal tapes in a very similar manner. Program 1 in Appendix B is an example of a program used to collect information at the HU level from the HH and crime records.

From the longitudinal tapes, we have selected a 1% subsample of HUs by choosing the 89th HU (a random starting place) and then selecting every 100th HU thereafter along with its corresponding individual and victimization information. Because the HUs on the Reiss tapes are ordered by their identification numbers, no two HUs from the same cluster are included in this subsample and we need not concern ourselves with cluster effects for this subsample. The analyses presented in subsequent chapters are based on this subsample which contains 1539 HUs.

1.6. Missing Data

Data that is missing from sample surveys can generally be classified into two types: "unit" missing data and "item" missing data. Unit missing data refers to entire sets of data that are missing for potential respondents. This type of missing data occurs, for instance, when an HH refuses to respond or has been rotated out of the sample. Item missing data occur when answers to some questions are missing in an otherwise completed interview. Here we focus on entire missing interviews as opposed to unanswered questions in interviews that actually took place. We note that if we were to consider all of the information obtained at an HU during its 7 (8 or 9) potential interviews, then a missing interview could be viewed as item missing data.

If, for a given year, we had all 12 months of victimization information for every HU in the survey, we could easily estimate θ using the number of crime-free HUs divided by the total number of HUs. Unfortunately many HUs do not contribute a full year's information during the year of interest. In the 1% sample from Reiss's longitudinal files, for each of the years 1973-1975, only about 50% of the HUs that were in the sample for at least part of the year have 12 months of victimization information. Some of the data are missing because the HU was rotated out of the sample before all the information for the year had been collected or because the HU was not rotated into the sample until after July of the year of interest and thus information concerning the beginning of the year was never collected. Another source of missing data is nonresponse. An interviewer cannot, for a variety of reasons, always obtain an interview at an HU that is in the sample. Even though an interview does not take place, some background information such as land use, size of city, etc. as well as a reason for the noninterview, is entered in the HH record and this record appears on the data tapes. The possible reasons for nonresponse are detailed in the next section.

At the HU level, in theory, the only two sources of missing data are the rotation scheme and HU nonresponse. Because of errors in the data, the matching procedure used by Reiss to create the longitudinal files was not always able to "properly" match all of the information for one HU. Thus there are HUs that should have been in the sample at a certain time, but for which no record (of an interview or a nonresponse) exists on the longitudinal tapes. Throughout this monograph, such missing data have been treated as though they were due to the rotation scheme. We expect that the errors that caused these problems have been substantially reduced since the start-up of the survey (Bureau of the Census, personal communication) and that attempts to match more current data in the preparation of longitudinal files will not suffer as much from this type of problem.

If we move from the HU level to the HH level, the missing data problem becomes slightly more complex. Suppose that the first HH in a particular HU has several interviews followed by a nonresponse. Suppose also that the nonresponse is followed by an interview but that a new HH has moved in. Because of the manner in which the NCS data has been collected and recorded, we do not know whether to attribute the nonresponse to the first HH, the second HH, or a fraction to each. This problem will be discussed in more detail in Chapter 6.

Finally, if we move from the HH level to the person level, the missing data problem described above becomes much more difficult. When an HH changes, all of the persons change and thus we have the same sort of problem in attributing nonresponses to persons. In addition, even though an HH may respond, some of the individuals within the HH may not respond. They may be unwilling to participate, temporarily away, difficult to contact, ill, they may have moved away from the HH, etc. An additional problem relating to missingness is the result of errors in the data. Persons are numbered sequentially within HHs and sometimes the same person is erroneously assigned a different number at a subsequent interview.

The problem of estimating the proportion of crime-free persons will not be discussed further here, but some of the ideas discussed in Chapter 6 for estimation at the HH level could be adapted to estimation at the individual level.

1.6.1. Types of Nonresponse

Although an HU is selected by the previously described sampling scheme, the interviewer is not always able to obtain an interview at that HU. When this situation arises, the interviewer reports this HU as having a nonresponse and specifies its type. There are 3 different types of HU nonresponse (cf. U.S Department of Justice,1978) and they are described in detail as follows:

Type A 1) no one is at home despite repeated visits

2) the entire household is temporarily away during all of the interview period

3) the household refuses to give any information

4) the unit cannot be reached due to impassable roads

5) interview is not conducted due to serious illness or death in the family

Type B 1) unit is vacant regular housing

2) unit is vacant and used for storage

3) unit is occupied by persons usually residing elsewhere

4) unit unfit for habitation or to be demolished

5) unit under construction and not ready for occupancy

6) unit temporarily converted to business or storage

7) address identifies an unoccupied tent or trailer site

8) permit granted, but construction not started

Type C 1) no address was listed on the sample line of the listing sheet

2) unit demolished by time of enumeration

3) house or trailer has been moved

4) unit converted to permanent business or storage

5) unit has been merged with another unit

Generally speaking, a Type A nonresponse occurs when the unit exists and there are people living there, but they are not interviewed, a Type B nonresponse occurs when the unit exists but there are no occupants, and a Type C nonresponse occurs when the unit does not exist.

1.6.2. The Magnitude of the Nonresponse

One of the problems with which we are confronted when trying to estimate θ is that, for many HUs, one or more of the interviews covering the year of interest is unavailable. A large proportion of these interviews is unavailable due to nonresponse. In order to assess the magnitude of the nonresponse data, each of the 1539 HU's in the 1% subsample has been categorized for each of the years 1973-75 as having no nonresponse months, at least one nonresponse month, or being out of sample for the entire year. HUs with at least one nonresponse month have been further subdivided into at least one of the following categories:

a. at least one month of type A nonresponse

b. at least one month of type B nonresponse

c. at least one month of type C nonresponse

u. at least one month of unknown type of nonresponse

Category u. arises because there are several HUs that had nonresponses for which the variable describing the type of nonresponse has been incorrectly coded or is missing. The results of this tabulation are shown in Table 3.

In 1973 there were 718 + 246 = 964 HUs in our 1% sample that were in sample for at least part of the year. Of these HUs, 718, or about 74%, had no nonresponse and thus we have about a 26% nonresponse rate. When the BJS reports nonresponse rates (U.S. Department of Justice, 1981d), type B and C

TABLE 3

HUs in 1% sample categorized
by types of nonresponse

		1973	1974	1975
no nonresponse		718	849	822
nonresponse		246	273	283
	a.	56	63	63
	b.	179	185	195
	c.	29	35	22
	u.	4	4	13
out of sample		575	417	434

nonresponse months are treated as out-of-sample months. Thus for our 1% sample for 1973, supposing that the nonresponses of unknown type are actually type A nonresponses, we would find a nonresponse rate of 60/(718 + 60) or approximately 8%. Unfortunately, as we will see in Chapter 5, HUs with type B and C noninterviews tend to have higher victimization rates than HUs with no nonresponse. Treating them as though they were out of sample is thus inappropriate, and leads to overestimates of our cheery indicators.

1.6.3. Ignorable Sampling and Nonresponse Mechanisms

The analysis of data sets from which some items are unavailable is often performed by ignoring the missing and nonsampled data. Rubin (1976)[1] and Little (1980) have developed conditions under which ignoring missing and nonsampled items leads to appropriate likelihood inferences. We describe their approach in the NCS context here.

[1] Dawid (1979) has pointed out that there is an error in this paper but this error does not affect the results discussed here.

Let

$$X_{ij} = \begin{cases} 1 & \text{if HU } i \text{ is crime-free in month } j \\ 0 & \text{if HU } i \text{ is victimized in month } j \end{cases}$$

for $i = 1,\ldots,H$ and $j = 1,\ldots,J$, where H is the total number of HUs in the population, and let $\mathbf{X} = (X_{ij})$. Note that $\mathbf{X}$ will not in general be fully observed. Corresponding to each X_{ij} define the sample indicator function

$$S_{ij} = \begin{cases} 1 & \text{if HU } i \text{ is sampled in month } j \\ 0 & \text{otherwise} \end{cases}$$

and let $\mathbf{S} = (S_{ij})$. Denote the set of X_{ij}'s that are sampled by $\mathbf{X}_s$ and its complement by $\mathbf{X}_s^c$. Suppose that $\mathbf{X}$ has a distribution $f(\mathbf{x}\,|\,\gamma)$ where γ is an unknown vector parameter, and that, conditional on $\mathbf{X}$, $\mathbf{S}$ has a distribution $f(\mathbf{s}\,|\,\mathbf{x};\phi)$ where ϕ is an unknown vector parameter. Then the joint distribution of $\mathbf{X}$ and $\mathbf{S}$, given γ and ϕ is

$$f(\mathbf{x},\mathbf{s}\,|\,\gamma,\phi) = f(\mathbf{x}\,|\,\gamma)f(\mathbf{s}\,|\,\mathbf{x};\phi)\ . \qquad (1\text{-}1)$$

Since some of the sampled items will be missing we need to define a response indicator function for each $X_{ij} \in \mathbf{X}_s$. Letting

$$R_{ij} = \begin{cases} 1 & \text{if } X_{ij} \text{ responds} \\ 0 & \text{if } X_{ij} \text{ does not respond} \end{cases}$$

and $\mathbf{R} = (R_{ij})$, we note that any particular outcome $\mathbf{r}$ of $\mathbf{R}$ partitions $\mathbf{X}_s$ into two parts. One part, call it $\mathbf{O}$, consists of the X_{ij}'s that can be observed, and the other, call it $\mathbf{M}$, consists of the X_{ij}'s that are missing due to nonresponse.

The observed data then consist of (s,r,o), and the nonsampled and missing values are contained in x_s^c and **m**, respectively.

Suppose that, conditional on **X** and **S**, **R** has a distribution $f(\mathbf{r} \mid \mathbf{x},\mathbf{s};\psi)$ indexed by an unknown vector parameter ψ. The joint distribution of **X**, **S**, and **R** is then found, using (1-1), to be

$$f(\mathbf{x},\mathbf{s},\mathbf{r} \mid \gamma, \phi, \psi) = f(\mathbf{x} \mid \gamma) f(\mathbf{s} \mid \mathbf{x};\phi) f(\mathbf{r} \mid \mathbf{x},\mathbf{s};\psi) \,, \tag{1-2}$$

and the distribution of the data, (**S**,**R**,**O**), is

$$\begin{aligned} f(\mathbf{o},\mathbf{s},\mathbf{r} \mid \gamma, \phi, \psi) &= \int f(\mathbf{x},\mathbf{s},\mathbf{r} \mid \gamma, \phi, \psi) d\mathbf{x}_s^c d\mathbf{m} \\ &= \int f(\mathbf{x} \mid \gamma) f(\mathbf{s} \mid \mathbf{x};\phi) f(\mathbf{r} \mid \mathbf{x},\mathbf{s};\psi) d\mathbf{x}_s^c d\mathbf{m} \,. \end{aligned} \tag{1-3}$$

In many problems the missing and nonsampled items are ignored, that is, inference is based on the distribution of **O** obtained by averaging $f(\mathbf{x} \mid \gamma)$ over the missing and nonsampled items:

$$f(\mathbf{o} \mid \gamma) = \int f(\mathbf{x} \mid \gamma) d\mathbf{x}_s^c d\mathbf{m} \,. \tag{1-4}$$

Rubin (1976) and Little (1980) give the following sufficient conditions under which likelihood based inferences using (1-3) are equivalent to those using (1-4):

1. For each value of ϕ,

$$f(\mathbf{s} \mid \mathbf{x};\phi) = f(\mathbf{s} \mid \mathbf{o},\mathbf{x}_s^c,\mathbf{m};\phi) \,, \tag{1-5}$$

   for all values of $\mathbf{x}_s^c$ and **m**. That is, the distribution of the sample indicator function does not depend on the values of the nonsampled or missing items.

2. For each value of ψ,

$$f(r\,|\,x,s;\psi) = f(r\,|\,o,x_s^c,m,s;\psi)\ , \tag{1-6}$$

for all values of x_s^c and **m**. That is, the distribution of the response indicator function does not depend on the values of the nonsampled or missing items. In the terminology of Rubin this condition is referred to as the data being *missing at random*.

3. The parameters γ and $(\phi\,,\psi)$ are *distinct* in the sense that the joint parameter space of γ and $(\phi\,,\psi)$ factorizes into the product of the parameter space of γ and the parameter space of $(\phi\,,\psi)$.

If these conditions hold, the sampling and nonresponse mechanisms are said to be ignorable and likelihood based inferences using (1-4) will be appropriate.

The following two examples are adapted from Rubin's paper and may help to clarify the "missing at random" and "ignorable sampling and nonresponse mechanisms" concepts.

Example 1: Suppose there are k persons in the population of interest and we attempt to determine, by means of a questionnaire, if each has or has not been the victim of a crime in the past six months. Suppose also that each person has constant probability, ψ, of failing to respond to the questionnaire. Then

$$f(s\,|\,x;\phi) = \prod_i I(s_i=1) \tag{1-7}$$

since all of the persons are sampled, and

$$f(r\,|\,x,s;\psi) = \prod_{i=1}^{k} \psi^{(1-r_i)}(1-\psi)^{r_i}\ . \tag{1-8}$$

Since this expression is independent of **x**, the data are always missing at random. ■

Example 2: Let x_i be the length of time since the last victimization for the ith subject in a victimization survey. Suppose that $r_i = 0$ if x_i is greater than ψ, which is the mean length of time since the last victimization. Thus we only measure x_i for subjects whose time since last victimization is less than average. Then,

$$f(s \mid \mathbf{x};\phi) = \prod_i \; I(s_i=1) \tag{1-9}$$

since all of the people are sampled, and

$$f(r \mid \mathbf{x},s;\psi) = \prod_i I\{I(x_i<\psi) = r_i\} \;. \tag{1-10}$$

Now

$$f(r \mid o,m,s;\psi) = \prod_{x_i \in o} I\{I(x_i<\psi) = r_i\} \prod_j I\{I(m_j<\psi) = r_i\} \tag{1-11}$$

which will be equal to (1-10) for all values of **m** only if there are no values in **m**, i.e. only if the time since the last victimization is observed for all of the subjects. Thus the data are missing at random only in the trivial case when none of the data are missing. ■

The complex random sampling scheme described earlier in this chapter assures that the sampling mechanism is ignorable. (We note that the sampling frame, that is, the list from which HUs are chosen, may contain inaccuracies causing certain subsets of the population to be under-represented. Using an ignorable sampling mechanism to chose units from such a frame would result in a sample that was not representative of the entire population. We wll not address problems of adjusting for any inadequacies of the sampling frame here.)

The assumption that the nonresponse mechanism is ignorable is not one which is easy to justify, especially in the NCS context. Certain subsets of the population (some of the same subsets that are under-represented in the sampling frame) tend to be nonrespondents more often than others. Nonetheless we begin by invoking this assumption in Chapters 2, 3, and 4 and examine its justification more closely in Chapter 5.

CHAPTER 2

SOME INTUITIVE PREVALENCE ESTIMATORS

In this chapter we discuss several ad hoc estimators of the prevalence of crime, including those that the BJS has used for the percentage of *households touched by crime* (U.S. Department of Justice, 1981a). These estimators are not derived from probabilistic models of victimization but rather are intuitive in nature. On the one hand, they have the advantage of being relatively easy to understand and straightforward to compute. On the other hand, it is difficult to determine to what extent they actually reflect the quantity that we are trying to estimate. Nonetheless, they are very helpful as a starting point for estimating θ and to pinpoint some difficulties in this estimation.

2.1. Two Ad Hoc Estimators

The first of the ad hoc estimators of θ, the probability that an HU is not victimized in a given year, is as follows:

1. Consider each HU that had an interview covering at least one month of the year under consideration.

2. If there is no victimization reported as having occurred in the year, consider the HU as being crime-free.

3. For each HU determine the number of interview months in the year (that is, the number of months covered by an interview).

4. Use as an estimator

$$\hat{\theta}_1 = \frac{\text{number of interview months for crime-free HU's}}{\text{total number of interview months}} . \qquad (2\text{-}1)$$

If an HU has reported a victimization in any month of the given year, then we are sure that it was not crime-free. Suppose that for such an HU we have 5 months of information, and that we could somehow obtain the missing 7 months of information. Regardless of how many of these 7 months were crime-free and how many were not, this HU would still contribute nothing to the numerator of $\hat{\theta}_1$ and 12 to the denominator. Thus, once we know that an HU has been victimized, the rest of the information is irrelevant and we may as well suppose that we have all 12 months of information for this HU. With this in mind we may prefer a variation of $\hat{\theta}_1$. If we let CFM be the number of interview months for crime-free HU's (the numerator of $\hat{\theta}_1$) then we can express the modified estimator as

$$\hat{\theta}'_1 = \frac{\text{CFM}}{12 \times \text{number of victimized HU's} + \text{CFM}} . \qquad (2\text{-}2)$$

Since the denominator of $\hat{\theta}'_1$ is larger than the denominator of $\hat{\theta}_1$, $\hat{\theta}'_1 \le \hat{\theta}_1$. Notice that, even though we are not using a model based approach, these estimators implicitly treat the nonresponse data as though they were missing at random, i.e. the nonresponse data and the data that are missing due to the rotation scheme are treated identically.

Using the 1% sample of 1539 HUs, described in Chapter 1, we have computed values for $\hat{\theta}'_1$ and $\hat{\theta}_1$ for 1973, 1974, and 1975. (Because of difficulties in the start-up of the survey and an irregular rotation scheme in 1972, we have not computed values of any of the estimators for that year, even though there are some data available.) The quantities used in computing $\hat{\theta}_1$ and $\hat{\theta}'_1$ are displayed in Table 4 as well as the estimates. The 4th and 5th lines of the table contain the number of HUs with 12 months of information for the year of interest and the total number of HUs that contribute at least 1 month of information for the given year, respectively. These two quantities were not used in the computations of the estimates, but from them we see that not all of the 1539 HUs in the sample are used in each estimate and that quite a few of the HUs that are used in the estimates do not contribute a full year's information.

TABLE 4

Ad Hoc Estimates

	1973	1974	1975
crime-free months	5774	6086	6244
interview months	8182	8461	8535
victimized HUs	231	248	228
HUs with 12 months	550	451	452
total HUs	836	966	955
$\hat{\theta}_1$	.708	.719	.732
$\hat{\theta}'_1$	.676	.672	.695

In order to get a rough idea of the possible sensitivity to nonresponse associated with the estimator $\hat{\theta}_1$, we have computed upper and lower bounds. The upper bound is obtained by assuming that every nonresponse month was a crime-free month and the lower bound was obtained by assuming that every nonresponse month was a month in which a victimization occurred. These bounds are shown in Table 5 along with the original estimates. Notice that

TABLE 5

Ad Hoc Estimates and Bounds

	1973	1974	1975
upper bound	.753	.760	.770
$\hat{\theta}_1$	.706	.719	.732
$\hat{\theta}'_1$	.676	.672	.695
lower bound	.497	.519	.523

the upper bounds are much closer to the actual values of $\hat{\theta}_1$ than are the lower bounds. This is due to the fact that the lower bounds were obtained by supposing that a victimization, which is a relatively uncommon event, occurred in each nonresponse month.

2.2. The BJS Estimators

The Bureau of Justice Statistics (1981a,1982) has published estimates of the proportion of *households touched by crime* (i.e., $1-\theta$) for each of the years 1975-1981. To calculate these estimates the Bureau of the Census (Alexander,1981) developed two ad hoc estimators whose form is similar to the estimators discussed above. BJS officials made a decision to publish estimated rates for 1980 before the end of March 1981. Since it would be necessary to have the information from interviews through June of 1981 in order to calculate 1980 rates, data for HUs that would have been interviewed after January of 1981 were imputed from their corresponding 1980 interviews as shown in Figure 1. For instance, for an HU in panel 3, information collected at the March 1980 and September 1980 interviews would be used as the information for 1980 although those 2 interviews actually cover the period of September 1979 through August 1980. Note that, in this way, it is possible

FIGURE 1

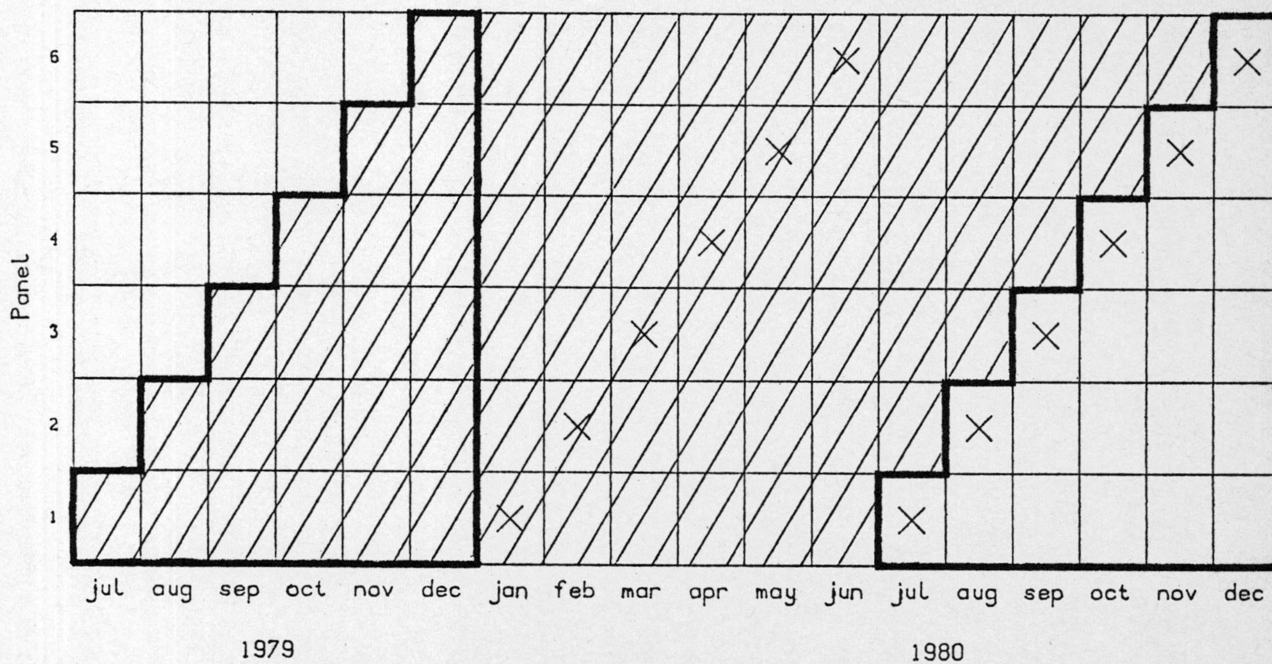

X indicates the month of interview.
Shaded area indicates months covered by those interviews.
Data for the interview months in the bordered section on the left are used as the values of the unavailable data for the interview months in the bordered section on the right.

to estimate the proportion of HUs victimized in a year by considering only two records for each HU as opposed to the three records which would normally be used for HUs in panels 2 through 6.

Because of the previously mentioned difficulties of estimating θ at the HH level, the BJS proportion of households touched by crime is actually based on HUs, not HHs. Thus the estimate is actually the proportion of housing units touched by crime. In order to calculate the BJS estimates, each HU is first classified according to the number and type of noninterview. (For a description of the types of noninterview, see Chapter 1.) The classifications of the HUs are as follows:

group a : both records are interviews

group b : only the first record is an interview--the second record is missing because the HU was rotated out of the sample, or the second interview is a type A noninterview.

group c : only the second record is an interview--the first record is missing because the HU had not yet been rotated into the sample or the first record was a type A noninterview.

group d : the first record is an interview, the second record is a type B or C noninterview

group e : the first record is a type B or C noninterview, the second is an interview

group f : neither record is an interview

From these groups, the following quantities are computed:

H_a = # of HUs in group a

H_b = # of HUs in group b

H_c = # of HUs in group c

H_d = # of HUs in group d

H_e = # of HUs in group e

C_a = # of HUs in group a that report at least 1 victimization in either interview

C_{a_1} = # of HUs in group a that report at least 1 victimization in the first interview

C_{a_2} = # of HUs in group a that report at least 1 victimization in the second interview

C_b = # of HUs in group b that report at least 1 victimization in the first interview

C_c = # of HUs in group c that report at least 1 victimization in the second interview

C_d = same as C_b but for group d

C_e = same as C_c but for group e

The BJS estimators are then given by

$$R_1 = \frac{2C_a + C_d + C_e + (C_b+C_c)\left(\frac{2C_a}{C_{a_1}+C_{a_2}}\right)}{2H_a + H_b + H_c + H_d + H_e} \qquad (2\text{-}3)$$

and

$$R_2 = \frac{2C_a + C_d + C_e + C_b\left(\frac{C_a}{C_{a_1}}\right) + C_c\left(\frac{C_a}{C_{a_2}}\right)}{2H_a + H_b + H_c + H_d + H_e} \qquad (2\text{-}4)$$

Each of these estimators scales the observed variables in order to account for the missing information for HUs in groups b and c (that is, for type A noninterviews and for HUs that are only in the sample for part of the year), but not for HUs in groups d and e (that is, for type B and C noninterviews). For a more detailed explanation the reasoning behind these estimators, see Alexander and Roebuck (1983). An example of the calculation of R_1 may help clarify the rationale behind the estimator. Consider the following hypothetical data:

	first record	second record	either
<u>group a</u>			
total HUs	$630=H_a$	$630=H_a$	$630=H_a$
victimized HUs	$90=C_{a_1}$	$80=C_{a_2}$	$134=C_a$
<u>groups b and c</u>			
total HUs	$120=H_b$	$130=H_c$	unknown
			use $125 = \frac{120+130}{2} = \frac{H_b+H_c}{2}$
victimized HUs	$15=C_b$	$18=C_c$	unknown
			use $26 = [15+18]\times\left[\frac{139}{90+80}\right]$ $= [C_b+C_c]\times\left[\frac{C_a}{C_{a_1}+C_{a_2}}\right]$
<u>groups d and e</u>			
total HUs	$30=H_d$	$20=H_e$	unknown
			use $25 = \frac{30+20}{2} = \frac{H_d+H_e}{2}$
victimized HUs	$5=C_d$	$3=C_e$	unknown
			use $4 = \frac{5+3}{2} = \frac{C_d+C_e}{2}$

Aggregating the above quantities yields

$$\text{total HUs} = 630 + 125 + 25 = 780$$

$$= H_a + \tfrac{1}{2}(H_b + H_c + H_d + H_e)\ ,$$

$$\text{total victimized HUs} = 134 + 26 + 4 = 164$$

$$= C_a + [C_b+C_c] \times \left[\frac{C_a}{C_{a_1}+C_{a_2}}\right] + \frac{C_d+C_e}{2}\ ,$$

and R_1 is just the ratio of total victimized HUs to total HUs. Note that the sum of the victimized HUs in groups b and c has been scaled down by the factor $C_a/(C_{a_1}+C_{a_2})$ which is calculated from group a to account for HUs that would have reported a victimization at both interviews had both interviews taken place. The sum of the victimized HUs in groups d and e is divided by 2 to reflect the fact that these HUs are only contributing one half of the year's data. It is assumed that no victimizations occurred at these HUs during the period covered by the missing record since the noninterview was a type B or C and hence no one was living in the unit (at least at the scheduled interview time). Since

$$1 \geq \frac{C_a}{C_{a_1}+C_{a_2}} \geq \frac{\max(C_{a_1},C_{a_2})}{C_{a_1}+C_{a_2}} \geq \frac{1}{2}, \tag{2-5}$$

the scaling factor for groups b and c is greater than or equal to the scaling factor for groups d and e. Thus HUs that have type A noninterviews or are out of sample for part of the year are weighted more heavily in R_1 than HUs that have had a type B or C noninterview.

The BJS estimator R_2 is very similar to R_1, the only difference being that separate scaling factors are used for groups b and c. The scaling factor for group b is computed using HUs from group a that reported a victimization only at the first interview and those that reported a victimization at both interviews. The scaling factor for group c is similar, using the HUs from group a that reported victimizations at the second interview only.

Instead of using actual counts in the above quantities, the Bureau of the Census uses the sample-based weights associated with the appropriate HUs. Thus, for instance, the $2H_a$ in the denominator of R_1 and R_2 is actually the total weight from both records of the group a HUs. The difficulty in general with using the sample-based weights in longitudinal analyses is that the weights change from interview to interview. Several HUs in our sample of 1539 exhibited substantial variation in these weights. For example, in one selected unit, the household weight varied from 964.996 to 1120.389, while the weight for one of the individuals in the HU varied from 874.662 to 1389.469. Because of this problem and others to be discussed in Chapter 3, the calculations and analysis presented here treat the above quantities as actual counts of HUs.

We have computed values of R_1 and R_2 from the 1% sample of 1539 HUs (using unweighted counts) and the results are shown in Table 6. As we saw in Table 3, there are several HUs for which the type of noninterview is unknown. By variously assigning these HUs to the different groups, we obtained the upper and lower limits on R_1 and R_2 that are displayed in Table 6. Note that the limits are very close to each other and thus the few HUs with unknown types of nonresponse do not have much effect on the estimates. Of the HUs with unknown nonresponse type for 1974, none were victimized and so the upper and lower bounds are the same for that year. For comparative purposes, we include the values of $1 - \hat{\theta}_1$ and $1 - \hat{\theta}'_1$ in the table.

From this table we see that the "BJS estimates" for 1974 and 1975 are very close to one another. The estimates for R_1 and R_2 for each of the years are also very close, with R_2 tending to be slightly larger. The BJS estimates are much closer to the comparable estimates obtained from $\hat{\theta}'_1$ than those obtained from $\hat{\theta}_1$. The BJS has reported its proportion of HUs touched by crime for the years 1975 through 1981. For 1975 the reported figure is .320 which differs from the values of R_1 and R_2 reported here because of at least the following three reasons:

1. The Bureau of the Census has done its own matching of records

while Reiss' matchings have been used here. If different criteria were used to determine whether or not two records matched, the resulting longitudinal files could be different. In addition, Reiss has tried to match all of the information for each HU from 1972 through 1975 while the Bureau of the Census only matched the two records necessary for the estimation for each year.

2. The numbers presented in this chapter were calculated from a *sample* of the full data set.

3. The rates reported by BJS are based on weighted counts whereas those given here are based strictly on counts.

TABLE 6

BJS Estimates

		1973	1974	1975
R_1	upper	.329	.311	.311
	lower	.327	.311	.311
R_2	upper	.333	.313	.313
	lower	.320	.313	.312
$1 - \hat{\theta}_1$		.292	.281	.268
$1 - \hat{\theta}'_1$		.324	.328	.305

In the next chapter we examine the ad hoc estimators and the BJS estimators, and attempt to find probabilistic models under which these estimators are sensible as estimators of θ.

CHAPTER 3

CONSISTENCY OF THE AD HOC MODELS

3.1. Modelling Victimization

Analyses of the NCS data performed to date by the Bureau of the Census and the BJS have generally used a finite population model. The values of the items for the units are treated as fixed but unknown and inferences are based on the distribution which results from taking a probability sample of units in the population. In particular, the victimization profile is considered as fixed and the only source of variance is due to the fact that we do not sample all of the HUs in the population. It is this perspective that led to the BJS estimators of θ described in Chapter 2. Alternatively, we can consider the victimization profile itself to be a random variable with a certain distribution and thus we have two sources of variation: the variance of this random variable and the variance due to the sampling. This kind of modelling is generally referred to in the sampling literature as superpopulation modelling. In this chapter we adopt the superpopulation approach and examine several possibilities for the distribution of the victimization random variable.

When taking a finite population sampling approach (e.g. see Kish (1965), Cochran (1977), or Hansen, Hurwitz, and Madow (1953a and b)) and estimating averages and totals, the individual sample values are weighted by the reciprocal of the probability of selection of that sample unit. The Bureau of the Census has carried this weighting approach further through the inclusion of additional factors. These six factors are:

1. a basic weight which is the reciprocal of the initially designed probability of selection.

2. a duplication weight. It sometimes happens that when an interviewer visits a particular location, he/she finds that there is actually more than one HU at that location. Sometimes only a fraction of these additional HUs will be interviewed. In this case, a duplication weight is needed to adjust the overall weight for these units.

3. a within HH noninterview weight .

4. a household noninterview weight. This adjustment puts more weight on HUs that are similar to the noninterviewed HH to compensate for the fact that no information is collected at the nonresponding unit. The noninterviewed HH is assumed to be like other HUs in the same rotation group, in nearby PSUs, with the same race of head of HH (white, non-white), type of PSU (SMSA, non-SMSA), and residence (central city, urban, rural farm, rural non-farm).

5. a first stage adjustment weight. This adjustment reduces the contribution to the variance of the non-self-representing PSU's, based on data from the 1970 Census. A different weight is computed for each race and residence combination.

6. a second stage ratio adjustment weight. This adjustment brings the sample figures into closer agreement with current Census estimates of population distribution as cross-classified by age, sex, and race.

The tables of NCS data that the U.S. Department of Justice publish all consist of weighted aggregates. Unfortunately, there is little theoretical justification for using these sample-based weights in analyses that involve superpopulation modelling. In fact, if we intend to use any of the quantities that were used to compute the weights as covariates in our model, then the use of these sample-based weights may be inappropriate. For this reason, the analyses in this thesis were performed using actual counts rather than the weighted counts.

In this chapter we "derive" probabilistic models under which the estimators described in the previous chapter are consistent. If we consider the housing units in the survey to be a random sample from an infinite superpopulation, then the consistency of an estimator as used here implies, roughly, that as the number of observations tends to infinity, the value of the estimator tends to the value of the parameter it estimates. For a further discussion of consistency, see Cox and Hinkley (1974). In particular, the ad hoc estimator of Chapter 2 is found to be consistent only under a model which assumes that the probability of reporting a victimization does not depend on the number of months of information that a housing unit contributes during the year of interest. The modified version of this estimator is found to be consistent under a model which seems, at least for a small sample, to fit the data well. The models under which the Bureau of Justice Statistics' estimators are consistent are similar to the model for the ad hoc estimator in that the probability of victimization in one half of the year must be the same as the probability of victimization in an entire year.

To derive these probabilistic models, we must make several assumptions. First we must assume that we can treat the nonresponse and nonsampled data in the same manner, or we must consider only HUs that have no nonresponse.

In addition, we must assume that HUs are victimized independently. This assumption may not hold in the entire data set since HUs are sampled in clusters, but since we have no HUs from the same cluster in our subsample the assumption seems reasonable for the present. Lastly, we assume that the probability that an HU does not report any crimes during the year of interest depends only on θ, the probability of being crime-free for the year, and the number of months of information that the HU contributes for that year.

3.2. Consistency of the Ad Hoc Estimator

Recall that the ad hoc estimator of θ discussed in Chapter 2 is

$$\hat{\theta}_1 = \frac{\text{\# of interview months in crime-free HUs}}{\text{\# of interview months}} . \tag{3-1}$$

We shall assume that there is no nonresponse and thus the only missing data are those due to the sample rotation scheme. Let n_i be the number of HUs that contribute exactly i interview months in the year of interest (i = 1, ..., 12) and let

$$Y_{ij} = \begin{cases} 1 \text{ if the } j^{th} \text{ HU that contributes exactly i interview months is crime-free} \\ 0 \text{ otherwise} \end{cases}$$

for j = 1, ..., n_i and i = 1, ..., 12 . Suppose that the Y_{ij} are independent random variables and that Y_{ij} has a Bernoulli distribution with parameter $f(i,\theta)$ where $f(i,\theta)$ is any function of i and θ. Thus $f(i,\theta)$ is the probability that an HU that contributes exactly i months of information will be crime-free in those i months. Note that the independence is between HUs and not between months within an HU.

Under these assumptions, Y_{i+} (the number of HUs that contribute exactly i interview months and are crime-free) has a binomial $[n_i, f(i,\theta)]$ distribution. Then, $\hat{\theta}_1$ can be written as

$$\hat{\theta}_1 = \frac{\sum_{i=1}^{12} iY_{i+}}{\sum_{i=1}^{12} in_i} = \frac{\sum_{i=1}^{12} in_i\overline{Y}_{i+}}{\sum_{i=1}^{12} in_i} \tag{3-2}$$

where $\overline{Y}_{i+} = Y_{i+}/n_i$. Letting $a_i = n_i/N$ (where $N = \sum_{i=1}^{12} n_i$) we have

$$\hat{\theta}_1 = \frac{\sum_{i=1}^{12} ia_i\overline{Y}_{i+}}{\sum_{i=1}^{12} ia_i}. \tag{3-3}$$

By letting $N \to \infty$ while the a_i remain constant (or assuming that as N increases each of the a_i's tends to a constant), and using the strong law of large numbers, we get as the limit

$$\lim_{N\to\infty} \frac{\sum_{i=1}^{12} ia_i\overline{Y}_{i+}}{\sum_{i=1}^{12} ia_i} = \frac{\sum_{i=1}^{12} ia_i f(i,\theta)}{\sum_{i=1}^{12} ia_i} \quad \text{(almost surely).} \tag{3-4}$$

If $\hat{\theta}_1$ is to be consistent, then $\hat{\theta}_1$ must converge in probability to θ. This, together with (3-4), implies that

$$\frac{\sum_{i=1}^{12} ia_i f(i,\theta)}{\sum_{i=1}^{12} ia_i} = \theta. \tag{3-5}$$

We would like to use this estimator ($\hat{\theta}_1$) for any $(a_1, a_2, \ldots, a_{12})$ that the design might specify, or at least for any $(a_1, a_2, \ldots, a_{12})$ in some neighborhood in the hyperplane given by $\sum_{i=1}^{12} a_i = 1$. In other words, we do not want the values of $f(i,\theta)$, the probability that an HU that contributes exactly i months of data is crime-free in those i months, to depend on the design. The following lemma will help us to find values of $f(i,\theta)$ that satisfy (3-5) for any $(a_1, a_2, \ldots, a_{12})$.

Lemma 1: Suppose that

$$\frac{\sum_{i=1}^{12} w_i a_i}{\sum_{i=1}^{12} z_i a_i} = \theta \qquad (3\text{-}6)$$

for all $(a_1, a_2, ..., a_{12})$ in some neighborhood on the hyperplane given by $\sum_{i=1}^{12} a_i = 1$. Then $w_i/z_i = \theta$ for $i = 1, ..., 12$.

Proof: Rewriting θ in terms of $a_1, ..., a_{11}$, we get

$$\frac{\sum_{i=1}^{11} w_i a_i + w_{12}(1 - \sum_{i=1}^{11} a_i)}{\sum_{i=1}^{11} z_i a_i + z_{12}(1 - \sum_{i=1}^{11} a_i)} = \theta \ . \qquad (3\text{-}7)$$

Letting ν represent the numerator of the above expression and δ the denominator and taking the derivative with respect to a_j, we get

$$\frac{\delta(w_j - w_{12}) - \nu(z_j - z_{12})}{\delta^2} = 0$$

$$\frac{w_j - w_{12}}{\delta} - \left(\frac{\nu}{\delta}\right) \frac{z_j - z_{12}}{\delta} = 0$$

$$\frac{(w_j - w_{12}) - \theta(z_j - z_{12})}{\delta} = 0$$

$$w_j - \theta z_j = w_{12} - \theta z_{12} \ . \qquad (3\text{-}8)$$

Next, let $c = w_{12} - \theta z_{12}$. Thus $w_j - \theta z_j = c$ and

$$\frac{\sum_{i=1}^{12} (c + \theta z_i) a_i}{\sum_{i=1}^{12} z_i a_i} = \theta \ . \qquad (3\text{-}9)$$

$$\frac{c + \theta \sum_{i=1}^{12} z_i a_i}{\sum_{i=1}^{12} z_i a_i} = \theta$$

$$\frac{c}{\sum_{i=1}^{12} z_i a_i} = 0$$

$c = 0$

Therefore $w_i - \theta z_i = 0$ and hence $w_i/z_i = \theta$. ∎

Applying Lemma 1 to expression (3-5) we find that

$$\frac{if(i,\theta)}{i} = \theta ,$$

or

$$f(i,\theta) = \theta \quad i = 1, \ldots, 12 . \tag{3-10}$$

Thus the only functions $f(i,\theta)$ which make $\hat{\theta}_1$ consistent for θ are $f(i,\theta) = \theta$. This implies that the probability that an HU that contributes exactly i months of information will be crime-free during those i months is constantly θ, regardless of i -- not a very reasonable model! Thus we have shown that, if we are willing to assume that HUs are victimized independently, the only model under which $\hat{\theta}_1$ is consistent is not sensible.

3.3. The Modified Version of the Ad Hoc Estimator

We can go through an argument similar to that in the previous section to find the functions $f(i,\theta)$ which make the modified version of the ad hoc estimator ($\hat{\theta}'_1$) consistent for θ. In the modified version of the ad hoc estimator, HUs that were victimized are treated as though they had contributed 12 months of information since, regardless of what information would have been gathered in the 12 - i months for which they were not in the sample, we would still treat the HU as a victimized HU. Thus the modified version of $\hat{\theta}_1$ is

$$\hat{\theta}'_1 = \frac{\text{\# interview months in crime-free HUs}}{12(\text{\# victimized HUs}) + (\text{\# interview months in crime-free HUs})} . \quad (3\text{-}11)$$

Using the same notation and assumptions as in the previous section, we have

$$\hat{\theta}'_1 = \frac{\sum_{i=1}^{12} i\alpha_i \overline{Y}_{i+}}{12[\sum_{i=1}^{12} \alpha_i(1-\overline{Y}_{i+})] + \sum_{i=1}^{12} i\alpha_i \overline{Y}_{i+}} . \quad (3\text{-}12)$$

Letting $N \to \infty$ as the α_i's remain constant (or assuming that each of the α_i's converges to a constant), and using the strong law of large numbers, we get as the limit

$$\lim_{N\to\infty} \hat{\theta}'_1 = \frac{\sum_{i=1}^{12} i\alpha_i f(i,\theta)}{12\sum_{i=1}^{12} \alpha_i[1-f(i,\theta)] + \sum_{i=1}^{12} i\alpha_i f(i,\theta)} \quad \text{(almost surely).} \quad (3\text{-}13)$$

If $\hat{\theta}'_1$ is to be consistent for θ, then $\hat{\theta}'_1$ must converge in probability to θ. This, along with (3-13), implies that

$$\frac{\sum_{i=1}^{12} i\alpha_i f(i,\theta)}{\sum_{i=1}^{12} \{12[1-f(i,\theta)] + if(i,\theta)\}\alpha_i} = \theta \quad (3\text{-}14)$$

where $\sum_{i=1}^{12} \alpha_i = 1$. Using Lemma 1 of the previous section, we have

$$\frac{if(i,\theta)}{12[1-f(i,\theta)] + if(i,\theta)} = \theta . \quad (3\text{-}15)$$

Then solving for $f(i,\theta)$, we find

$$f(i,\theta) = \frac{12\theta}{(12-i)\theta + i} . \quad (3\text{-}16)$$

Thus $\hat{\theta}'_1$ is consistent for θ under the model where the Y_{ij} are independent and

Y_{ij} has a Bernoulli distribution with parameter $12\theta/[(12-i)\theta + i]$. If the parameter takes any other form, $\hat{\theta}'_1$ will not be consistent under this independent Bernoulli model. From equation (3-13) we see that under this model $\hat{\theta}'_1$ is in fact strongly consistent for θ.

Thus, if we believe that $12\theta/[(12-i)\theta + i]$ is a reasonable form of the parameter, we have found a model under which $\hat{\theta}'_1$ is a reasonable estimator. Using this model, we can compute variance estimates for $\hat{\theta}'_1$. Let

$$S_1 = N^{1/2}[\sum_1^{12} i\alpha_i\bar{Y}_{i+} - \sum_1^{12} i\alpha_i f(i,\theta)] \tag{3-17}$$

and

$$S_2 = N^{1/2}[\sum_1^{12} (12-i)\alpha_i\bar{Y}_{i+} - \sum_1^{12} (12-i)\alpha_i f(i,\theta)] . \tag{3-18}$$

Since $N^{1/2}[\alpha_i\bar{Y}_{i+} - \alpha_i f(i,\theta)]$ has an asymptotic normal distribution with mean 0 and variance $\alpha_i f(i,\theta)[1-f(i,\theta)]$ and since the $\bar{Y}_{i+}$'s are independent, we find that S_1 and S_2 have an asymptotic bivariate normal distribution with mean 0 and variance

$$\begin{pmatrix} \sum_1^{12} i^2\alpha_i f(i,\theta)(1-f(i,\theta) & \sum_1^{12} i(12-i)\alpha_i f(i,\theta)(1-f(i,\theta) \\ \sum_1^{12} i(12-i)\alpha_i f(i,\theta)(1-f(i,\theta) & \sum_1^{12} (12-i)^2\alpha_i f(i,\theta)(1-f(i,\theta) \end{pmatrix} . \tag{3-19}$$

$\hat{\theta}'_1$ can be expressed as

$$\frac{\sum_1^{12} i\alpha_i\bar{Y}_{i+}}{12-\sum_1^{12}(12-i)\alpha_i\bar{Y}_{i+}} \tag{3-20}$$

and thus, by a straightforward application of the δ method (Bishop et al.,1975), we find that $N^{1/2}(\hat{\theta}'_1-\theta)$ has an asymptotic normal distribution with variance

$$\frac{1}{[12-\sum_1^{12}(12-i)\alpha_i f(i,\theta)]^4}\Big([12-\sum_1^{12}(12-i)\alpha_i f(i,\theta)]^2\ [\sum_1^{12} i^2\alpha_i f(i,\theta)(1-f(i,\theta))]\ +$$

$$2[12-\sum_1^{12}(12-i)\alpha_i f(i,\theta)]\ [\sum_1^{12} i\alpha_i f(i,\theta)]\ [\sum_1^{12} i(12-i)\alpha_i f(i,\theta)(1-f(i,\theta))]\ +$$

$$[\sum_1^{12} i\alpha_i f(i,\theta)]^2\ [\sum_1^{12}(12-i)^2\alpha_i f(i,\theta)(1-f(i,\theta))]\Big)\ . \tag{3-21}$$

By substituting $\hat{\theta}'_1$ for θ in expression (3-21), we can estimate the variance of $\hat{\theta}'_1$. The estimated standard deviations are displayed in Table 7.

We might now want to know if $\hat{\theta}'_1$ is also a maximum likelihood estimator under this model. The likelihood of θ given the data is

$$\text{lik}(\theta\,|\,y) = \prod_{i=1}^{12}\prod_{j=1}^{n_i}\left(\frac{12\theta}{(12-i)\theta+i}\right)^{y_{ij}}\left(\frac{i(1-\theta)}{(12-i)\theta+i}\right)^{1-y_{ij}}$$

$$= \prod_{i=1}^{12}\left(\frac{12\theta}{(12-i)\theta+i}\right)^{y_{i+}}\left(\frac{i(1-\theta)}{(12-i)\theta+i}\right)^{n_i-y_{i+}}\ . \tag{3-22}$$

Taking logarithms we find

$$\text{loglik}(\theta\,|\,y) = \sum_{i=1}^{12}\{y_{i+}\log 12\theta - y_{i+}\log[(12-i)\theta+i] + (n_i-y_{i+})\log[i(1-\theta)] - (n_i-y_{i+})\log[(12-i)\theta+i]\}$$

$$= \sum_{i=1}^{12}\{y_{i+}\log 12\theta + (n_i-y_{i+})\log[i(1-\theta)] - n_i\log[(12-i)\theta+i]\}\ . \tag{3-23}$$

Taking derivatives with respect to θ yields

$$\sum_{i=1}^{12}\left[\frac{y_{i+}}{\theta} + (n_i-y_{i+})\left(\frac{-1}{1-\theta}\right) - n_i\,\frac{12-i}{(12-i)\theta+i}\right]$$

$$= \frac{y_{++}}{\theta} - \frac{N-y_{++}}{1-\theta} - \sum_{i=1}^{12} n_i\,\frac{12-i}{(12-i)\theta+i}$$

$$= \frac{y_{++}-N\theta}{\theta(1-\theta)} - \sum_{i=1}^{12}\frac{n_i(12-i)}{(12-i)\theta+i}\ . \tag{3-24}$$

Thus the MLE, which will be denoted $\hat{\theta}_2$, is specified by

$$\frac{y_{++} - N\hat{\theta}_2}{\hat{\theta}_2(1-\hat{\theta}_2)} = \sum_{i=1}^{12} \frac{n_i(12-i)}{(12-i)\hat{\theta}_2 + i} . \tag{3-25}$$

Note that expression (3-25) depends on the data only through y_{++}, the total number of crime-free HUs, and not on the number of months of information that they contribute, that is, not on the y_{i+} as we might have suspected.

In general, we would have to solve for $\hat{\theta}_2$ using some iterative procedure, but in a few special cases, expression (3-25) can be solved explicitly as seen in the following examples.

Example 1: Suppose $n_1 = ... = n_{11} = 0$ and thus $n_{12} = N$. In this case we only observe data in full years. Then (3-25) becomes

$$\frac{y_{++} - N\hat{\theta}_2}{\hat{\theta}_2(1-\hat{\theta}_2)} = 0 \tag{3-26}$$

and

$$\hat{\theta}_2 = \frac{y_{++}}{N} . \qquad \blacksquare$$

Example 2: Suppose $n_j = N$ and $n_i = 0$ for $i \neq j$. Note that this is the only case where we can expect $\hat{\theta}_2$ to equal $\hat{\theta}'_1$ since $\hat{\theta}'_1$ depends on the y_{i+} and $\hat{\theta}_2$ depends only on y_{++}. Rewriting expression (3-25) we get

$$\frac{y_{++} - N\hat{\theta}_2}{\hat{\theta}_2(1-\hat{\theta}_2)} = \frac{N(12-j)}{(12-j)\hat{\theta}_2 + j} \tag{3-27}$$

or

$$y_{++}(12-j)\hat{\theta}_2 + y_{++}j - N\hat{\theta}_2^{\,2}(12-j) - Nj\hat{\theta}_2 = N(\hat{\theta}_2 - \hat{\theta}_2^{\,2})(12-j)$$

or

$$y_{++}(12-j)\hat{\theta}_2 + y_{++}j = 12N\hat{\theta}_2$$

or

$$\hat{\theta}_2 = \frac{jy_{++}}{12N - (12-j)y_{++}},$$

which is $\hat{\theta}'_1$. Thus $\hat{\theta}_2$ and $\hat{\theta}'_1$ agree in this special case. ■

Example 3: Suppose $n_6>0$, $n_{12}>0$, and $n_i=0$ for $i\neq 6$ or 12. Thus we observe full years and half years of data. This is the form in which the BJS uses the data in computing R_1 and R_2 as discussed in Chapter 2. In this case, expression (3-25) becomes

$$\frac{y_{++} - N\hat{\theta}_2}{\hat{\theta}_2(1-\hat{\theta}_2)} = \frac{6n_6}{6\hat{\theta}_2 + 6}. \tag{3-28}$$

This equation can be rewritten as

$$(N-n_6)\hat{\theta}_2^{\,2} + (N+n_6-y_{++})\hat{\theta}_2 - y_{++} = 0 ,$$

or

$$n_{12}\hat{\theta}_2^{\,2} + (n_{12}+2n_6-y_{++})\hat{\theta} - y_{++} = 0 .$$

Thus

$$\hat{\theta}_2 = \frac{-(n_{12}+2n_6-y_{++}) \pm \sqrt{(n_{12}+2n_6-y_{++})^2+4n_{12}y_{++}}}{2n_{12}}. \tag{3-29}$$

Note that:

1. The quantity under the square root is positive.

2. Since $\sqrt{(n_{12}+2n_6-y_{++})^2 + 4n_{12}y_{++}}$ is greater than or equal to $n_{12}+2n_6-y_{++}$ and since $n_{12}+2n_6-y_{++}$ is positive, the numerator of $\hat{\theta}_2$ will be nonnegative iff we add rather than subtract at the ± sign. Since the denominator is always positive, $\hat{\theta}_2$ will be nonnegative if we replace the ± sign with a + sign. Thus

$$\hat{\theta}_2 = \frac{-(n_{12}+2n_6-y_{++}) + \sqrt{(n_{12}+2n_6-y_{++})^2+4n_{12}y_{++}}}{2n_{12}}. \tag{3-30}$$

3. When $y_{++}=0$, $\hat{\theta}_2 = 0$ and when $y_{++} = n_6 + n_{12}$ (=N), $\hat{\theta}_2 = 1$.

4. We can show that the first derivative of $\hat{\theta}_2$ is positive and thus $\hat{\theta}_2$ increases in y_{++}. Hence $\hat{\theta}_2$ as given by equation (3-30) is real-valued and lies in the interval [0,1].

5. Furthermore the second derivative of $\hat{\theta}_2$ is positive. Figure 2 shows $\hat{\theta}_2$ plotted as a function of y_{++}. This curve lies below the line $\hat{\theta}_2 = y_{++}/N$. Thus $\hat{\theta}_2$ is always less than or equal to the observed proportion of crime-free HUs, a reasonable result since some of those HUs would have reported a victimization if we would have been able to obtain the full year's data for them. ■

Thus we see that the MLE of θ for the model of independence between HUs and $f(i,\theta) = 12\theta/[(12-i)\theta + i]$ is not in general equal to $\hat{\theta}'_1$. In some cases an explicit solution for $\hat{\theta}_2$ can be found. Otherwise it must be evaluated using some iterative procedure which we have done for each of the years 1973-1975. In the present analysis, only HUs for which there was no nonresponse during the year of interest are included. In Chapter 5, some of the estimates are computed for the entire 1% sample.

Evaluating the maximum likelihood estimator, $\hat{\theta}_2$, specified by equation (3-25) requires the use of an iterative procedure. The MINPACK package, written by B. S. Garbow, K. E. Hillstrom, and J. J. More of the Argonne National Laboratory, includes several programs to find a zero of a system of N non-linear equations in N variables by a modification of the Powell hybrid method. We used the HYBRD1 program, which estimates the Jacobian by a forward-difference approximation, to obtain the three values of θ given in Table 7. The program took about 0.5 seconds of CPU time on a DEC20 to compute all

FIGURE 2

The MLE, $\hat{\theta}_2$, versus
y_{++}, the total number of crime-free HUs

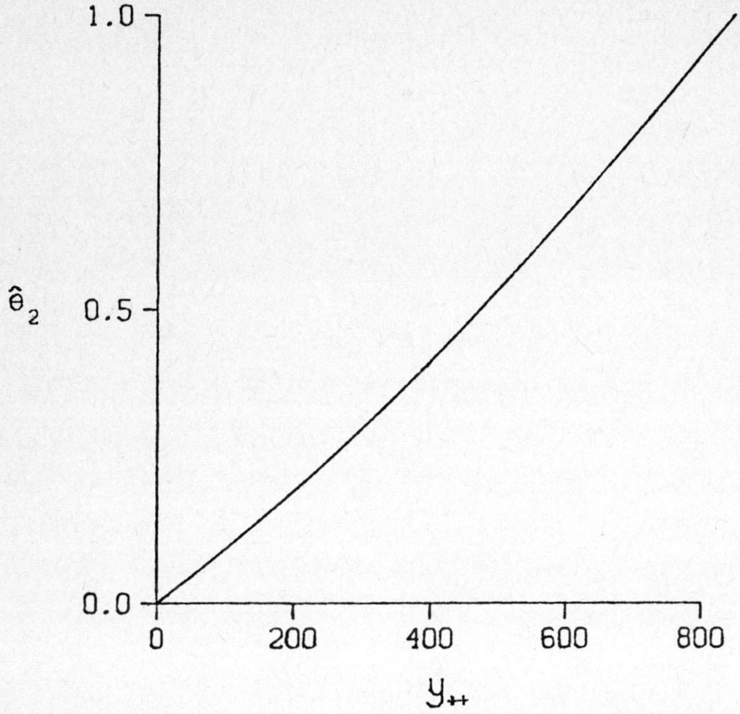

three of these values. Although we could have used the HYBRJ1 program, in which the user specifies the Jacobian, the rapid convergence of the HYBRD1 program indicated that the estimated Jacobian was adequate. Notice that, as expected, the values of $\hat{\theta}_2$ are close to the corresponding values of $\hat{\theta}'_1$ which have been computed for the HUs with no nonresponse and are given in Table 7.

TABLE 7

$\hat{\theta}'_1$, $\hat{\theta}_2$ and Estimated Standard Deviations

year	$\hat{\theta}'_1$	est. st.dev.	$\hat{\theta}_2$	est. st.dev.
1973	.694	.0342	.692	.0183
1974	.685	.0176	.682	.0099
1975	.685	.0280	.716	.0067

The log likelihood for the 1973 data is plotted in Figure 3 as a function of θ. It is clearly unimodal and so the result of the iterative procedure is determination of the global maximum. The plot seems to be rather flat near the maximum, possibly indicating a large variance for $\hat{\theta}_2$, but the log likelihood can be misleading in this respect. By adding the value of the log likelihood at $\hat{\theta}_2$ to the log likelihood and then exponentiating, we can compute a multiple of the likelihood of θ. (Note that by simply exponentiating the log likelihood we would have numbers that were too small to be handled by the computer.) This relative likelihood is plotted as a function of θ in Figure 4. We see that the likelihood actually has a fairly sharp peak despite the apparent flatness of the log likelihood.

To get a variance estimate for $\hat{\theta}_2$ under this model, we can calculate negative one times the second derivative of the log likelihood as an estimate of the information and use the reciprocal of this estimated information,

FIGURE 3

The log likelihood of θ
under the consistency model.
1973 data

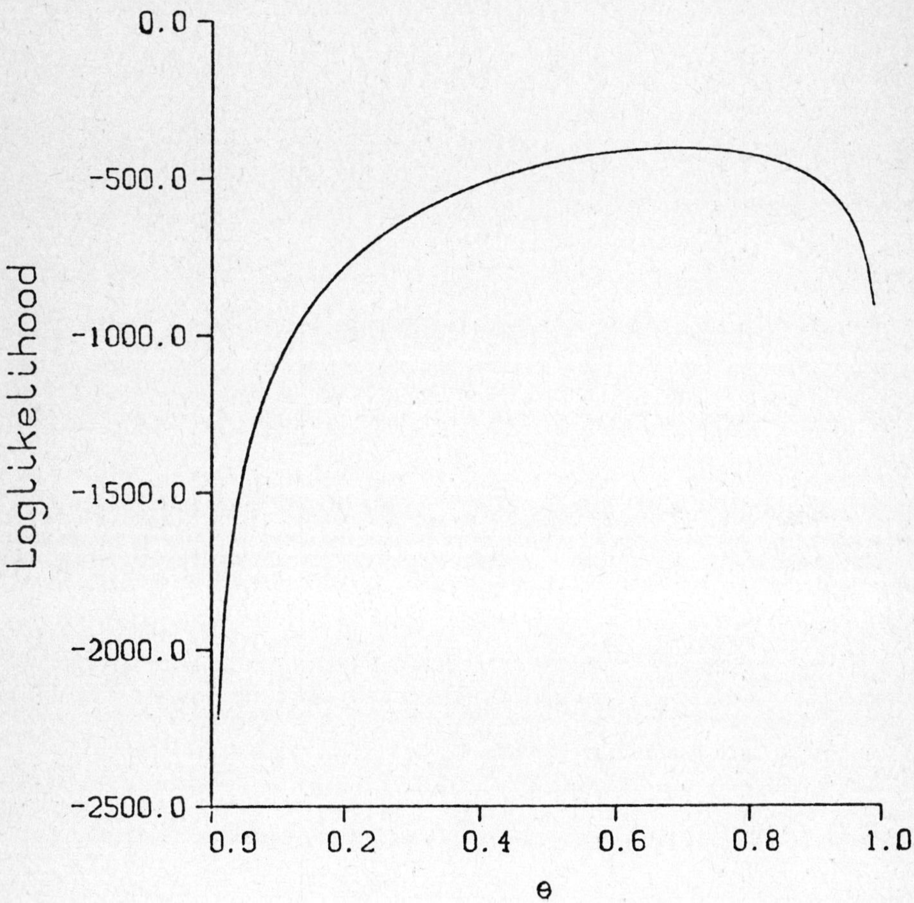

FIGURE 4

The relative likelihood of θ
under the consistency model.
1973 data

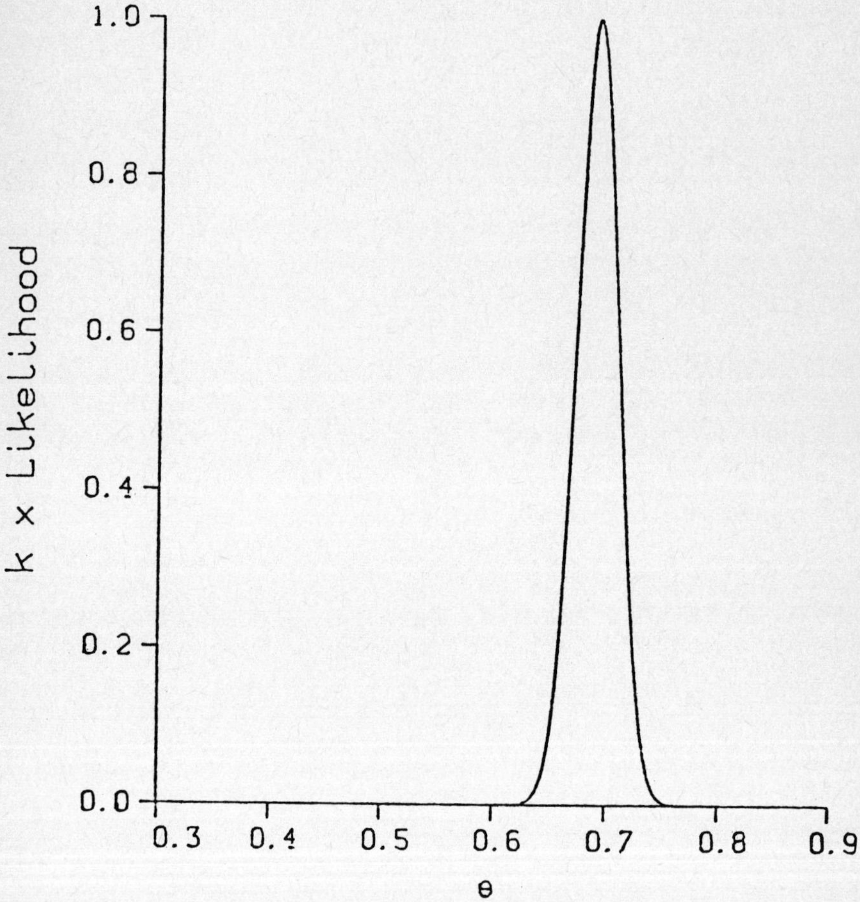

evaluated at $\hat{\theta}_2$, as an estimate of the variance. The log likelihood is given in equation (3-23) and taking the second derivative yields

$$\frac{d^2 \text{loglik}}{d\theta^2} = - \frac{N\theta(1-\theta) + (y_{++}-N\theta)(1-2\theta)}{\theta^2(1-\theta)^2} + \sum_{i=1}^{12} \frac{n_i(12-i)^2}{[(12-i)\theta+i]^2} \quad . \tag{3-31}$$

Thus an estimate of the variance is minus one over the right-hand side of equation (3-31) evaluated at $\hat{\theta}_2$. The results of these computations are shown in Table 7. We find that there does not seem to be much difference between the estimates of θ for 1973 and 1974 while the 1975 indicator appears to be larger than that for 1974.

The fit of this model can be tested by means of a χ^2 goodness-of-fit statistic. This statistic has been calculated for each of the years 1973-1975 again using only those HUs for which there was no nonresponse in the year of interest. As discussed above, the MLEs were evaluated using the program HYBRD1 from the MINPACK package. The results of these computations are shown in Table 8. We see that, since we have 10 degrees of freedom, the fit of the model to the data from each of the three years is fairly good.

TABLE 8

Fit of model $f(i,\theta) = 12\theta/[(12-i)\theta + i]$

1973 $\hat{\theta}'_1 = 0.694$ $\hat{\theta}_2 = 0.692$

months (i)	#HUs	#vict HUs	#cf HUs	$f(i,\theta)$	expected #cf HUs	residual	contribution to χ^2
1	19	2	17	.964	18.322	-1.322	.095
2	22	2	20	.931	20.483	-0.483	.011
3	15	0	15	.900	13.500	1.500	.167
4	19	3	16	.871	16.549	-0.549	.018
5	19	3	16	.844	16.032	-0.032	.000
6	29	7	22	.818	23.728	-1.728	.126
7	6	2	4	.794	4.765	-0.765	.123
8	4	1	3	.771	3.086	-0.086	.002
9	8	3	5	.750	6.000	-1.000	.167
10	5	2	3	.730	3.649	-0.649	.115
11	10	6	4	.711	7.105	-3.105	1.357
12	550	161	389	.692	380.782	8.218	.177

$\chi^2 = 2.359$

1974 $\hat{\theta}'_1 = 0.685$ $\hat{\theta}_2 = 0.682$

months (i)	#HUs	#vict HUs	#cf HUs	$f(i,\theta)$	expected #cf HUs	residual	contribution to χ^2
1	36	1	35	.963	34.656	0.346	.003
2	39	4	35	.928	36.188	-1.188	.039
3	30	7	23	.896	26.868	-3.868	.557
4	30	1	29	.865	25.965	3.035	.355
5	42	10	32	.837	35.168	-3.168	.285
6	41	8	33	.811	33.249	-0.249	.002
7	43	9	34	.786	33.805	0.195	.001
8	31	14	17	.763	23.649	-6.649	1.869
9	33	6	27	.741	24.450	2.550	.266
10	35	15	20	.720	25.206	-5.206	1.075
11	26	8	18	.701	18.215	-0.215	.003
12	451	129	322	.682	307.584	14.416	.676

$\chi^2 = 5.131$

continued

1975 $\hat{\theta}'_1 = 0.685$ $\hat{\theta}_2 = 0.716$

months (i)	#HUs	#vict HUs	#cf HUs	f(i,θ)	expected #cf HUs	residual	contribution to χ^2
1	24	2	22	.968	23.233	-1.233	.065
2	40	2	38	.938	37.521	0.479	.006
3	30	5	25	.910	27.295	-2.295	.193
4	39	4	35	.883	34.448	0.552	.009
5	30	5	25	.858	25.747	-0.747	.022
6	37	5	32	.835	30.880	1.120	.041
7	21	4	17	.812	17.056	-0.056	.000
8	39	7	32	.791	30.848	1.152	.043
9	34	3	31	.771	26.208	5.792	.876
10	33	10	23	.752	24.806	-1.806	.131
11	29	6	23	.733	21.271	1.792	.141
12	452	132	320	.716	323.688	-3.688	.042

$\chi^2 = 1.569$

Thus we see that the model under which the modified version of the ad hoc estimator is consistent does not require that the probability of reporting a victimization be independent of time in sample, and our 1% sample seems to fit this model fairly well.

Schervish (1983) has used a beta-binomial model to describe victimization. In this model the number of months of victimization for a given HU is assumed to have a binomial distribution with an unknown parameter p. This parameter, p, is then assumed to have a beta distribution with parameters α and β. Thus the victimizations for a particular HU are correlated with each other, but this correlation varies from HU to HU. Schervish has compared the results of the fit of his beta-binomial model with the results shown in Table 8 and has found that the estimates of θ, the expected number of victimized HUs, the residuals, and the fits for the three years of data are remarkably close for these two models. This is especially surprising since the beta-binomial model has two parameters whereas the model for which the modified version of the ad hoc estimator is consistent has only one. In addition, the beta-binomial estimates depend on the numbers of HUs contributing j months of information that have reported k months of victimization for all j from 1 through 12 and k from 0 to j as opposed to the one statistic, y_{++}, used in the computations for Table 8. Although it is more difficult to produce estimates of θ using the beta-binomial model, we note that this model does have a very intuitive interpretation.

3.4. Consistency of the BJS Estimators

In this section, we apply a consistency argument, similar to the one we used for the ad hoc estimators, to the BJS estimators which were discussed in Chapter 2. In order to do this, we first need to develop some additional notation. Let

$$Y_{a_1 j} = \begin{cases} 1 & \text{if } j^{th} \text{ HU in group a reports a victimization at the first interview} \\ 0 & \text{otherwise} \end{cases}$$

$$Y_{a_2 j} = \begin{cases} 1 & \text{if } j^{th} \text{ HU in group a reports a victimization at the second interview} \\ 0 & \text{otherwise} \end{cases}$$

$$Y_{bj} = \begin{cases} 1 & \text{if } j^{th} \text{ HU in group b reports a victimization at the first interview} \\ 0 & \text{otherwise} \end{cases}$$

and define Y_{cj}, Y_{dj}, Y_{ej} similarly. Then the quantities used to calculate R_1 are

$$C_a = \sum_{j=1}^{Ha} I(Y_{a_1 j} + Y_{a_2 j} > 0)$$

$$C_{a_1} = \sum_{j=1}^{Ha} Y_{a_1 j} \qquad C_{a_2} = \sum_{j=1}^{Ha} Y_{a_2 j}$$

$$C_b = \sum_{j=1}^{Hb} Y_{bj}$$

$$C_c = \sum_{j=1}^{Hc} Y_{cj}$$

$$C_d = \sum_{j=1}^{Hd} Y_{dj}$$

$$C_e = \sum_{j=1}^{He} Y_{ej}$$

R_1 is then given by

$$R_1 = \frac{2\sum_{j=1}^{Ha} I(Y_{a_1 j}+Y_{a_2 j}>0) + \sum_{j=1}^{Hd} Y_{dj} + \sum_{j=1}^{He} Y_{ej} + \left(\sum_{j=1}^{Hb} Y_{bj} + \sum_{j=1}^{Hc} Y_{cj}\right)\left[\dfrac{2\sum_{j=1}^{Ha} I(Y_{a_1 j}+Y_{a_2 j}>0)}{\sum_{j=1}^{Ha} Y_{a_1 j} + \sum_{j=1}^{Ha} Y_{a_2 j}}\right]}{2H_a + H_b + H_c + H_d + H_e} . \tag{3-32}$$

For each i = b, c, d, e, suppose that

$$Y_{ij} \sim \text{iid Bernoulli}[g(i,\theta)],\ j = 1, \ldots, H_i.$$

Thus $g(i,\theta)$ is the probability that an HU in group i reports a victimization. Suppose also that $Y_{a_1 j}$ and $Y_{a_2 j}$ have the following marginal distributions

$$Y_{a_1 j} \sim \text{iid Bernoulli}[g_1(a,\theta)],\ j=1, \ldots, H_a$$

and

$$Y_{a_2 j} \sim \text{iid Bernoulli}[g_2(a,\theta)],\ j=1, \ldots, H_a .$$

Note that the joint distribution of $Y_{a_1 j}$ and $X_{a_2 j}$ is then

		$Y_{a_2 j}$ = 0	$Y_{a_2 j}$ = 1	
$Y_{a_1 j}$	0	θ	$1-g_1-\theta$	$1-g_1$
	1	$1-g_2-\theta$	$g_1+g_2+\theta-1$	g_1
		$1-g_2$	g_2	

(3-33)

Letting $N = 2H_a + H_b + H_c + H_d + H_e$, we can rewrite R_1 as

$$R_1 = \frac{2H_a}{N}\left(\frac{\sum_{j=1}^{Ha} I(Y_{a_1 j}+Y_{a_2 j}>0)}{H_a}\right) + \frac{H_d}{N}\left(\frac{\sum_{j=1}^{Hd} Y_{dj}}{H_d}\right) + \frac{H_e}{N}\left(\frac{\sum_{j=1}^{He} Y_{ej}}{H_e}\right)$$

$$+\left[\frac{H_b}{N}\left(\frac{\sum_{j=1}^{Hb} Y_{bj}}{H_b}\right)+\frac{H_c}{N}\left(\frac{\sum_{j=1}^{Hc} Y_{cj}}{H_c}\right)\right]\left[\frac{2\{\sum_{j=1}^{Ha} I(y_{a_1j}+Y_{a_2j}>0)\}/H_a}{\sum_{j=1}^{Ha} Y_{a_1j}/H_a+\sum_{j=1}^{Ha} Y_{a_2j}/H_a}\right]. \tag{3-34}$$

Let $\alpha_a = 2H_a/N$ and $\alpha_i = H_i/N$ for i = b, c, d, e. Then $\sum_{i=a}^{e} \alpha_i = 1$ and if we hold the α_i's constant as N tends to ∞ (or if we assume that the α_i's each converge to a constant) we find that

$$\lim_{N\to\infty} R_1 = \alpha_a(1-\theta) + \alpha_d g(d,\theta) + \alpha_e g(e,\theta)$$

$$+ [\alpha_b g(b,\theta) + \alpha_c g(c,\theta)]\left[\frac{2(1-\theta)}{g_1(a,\theta) + g_2(a,\theta)}\right] \text{ (almost surely).} \tag{3-35}$$

In order for R_1 to be consistent, expression (3-35) must equal $1-\theta$. To find the forms of $g(i,\theta)$, i=a, ..., e for which (3-35) is equal to $1-\theta$, we need the following lemma.

Lemma 2: If $\sum_{i=a}^{e} w_i\alpha_i = 1-\theta$ for all $(\alpha_a, \ldots, \alpha_e)$ in some neighborhood in the hyperplane specified by $\sum_{i=a}^{e} \alpha_i = 1$, then $w_i = 1-\theta$, i=a, ..., e.

Proof: Rewriting the expression in terms of $\alpha_1, \ldots, \alpha_{11}$ we find that

$$\sum_{i=a}^{d} w_i\alpha_i + w_e(1 - \sum_{i=a}^{d} \alpha_i) = 1 - \theta\ . \tag{3-36}$$

Taking derivatives with respect to α_k yields

$$w_k + w_e(-1) = 0$$

$$w_k = w_e\ .$$

Thus $w_a = w_b = \ldots = w_e = w$, and

$$\sum_{i=a}^{e} w\alpha_i = 1 - \theta$$

$$w\sum_{i=a}^{e} \alpha_i = 1 - \theta$$

$w = 1 - \theta$. ∎

Applying Lemma 2 to equation (3-35), we have

i) $g(d,\theta) = 1 - \theta$,

ii) $g(e,\theta) = 1 - \theta$,

iii) $\dfrac{2g(b,\theta)(1-\theta)}{g_1(a,\theta) + g_2(a,\theta)} = 1 - \theta$, and thus $g(b,\theta) = \dfrac{g_1(a,\theta) + g_2(a,\theta)}{2}$,

iv) $g(c,\theta) = \dfrac{g_1(a,\theta) + g_2(a,\theta)}{2}$.

The restrictions on $g(d,\theta)$ and $g(e,\theta)$ are clearly unreasonable at least from an intuitive point of view. The probability that an HU that is interviewed once and is then demolished reports a victimization at that interview should not be the same as the probability that an HU that is interviewed twice reports a victimization at either of those interviews.

Referring to the hypothetical example in Chapter 2, we note that 134/630 seems to be a reasonable estimate of $1 - \theta$ and so does 26/125 . But 4/25 = $(C_d+C_e)/(H_d+H_e)$ seems to estimate the probability of being victimized in one half of the year and it is not being combined with the previous two quantities in a way that reflects this fact. The problem here is that we need to model the relation between the probability of being crime-free in half a year and the probability of being crime-free in a whole year, i.e. θ, if we are to be able to use the data from half of the year in an estimate of θ.

In a similar analysis for R_2, we find that restrictions i) and ii) remain unchanged while iii) and iv) become $g(b,\theta) = g_1(a,\theta)$ and $g(c,\theta) = g_2(a,\theta)$, respectively. Thus in either case we have a restriction which is intuitively unreasonable. This same problem exists for the estimator RNEW, a modification of the BJS estimators, suggested by Griffin (1981). In practice, this problem may not greatly affect the numerical results since the HUs in groups d and e are those which have had either a type B or C noninterview and there are few of these relative to the total number of HUs. Still, the impact of restrictions i) and ii) needs to be carefully examined by those who wish to use R_1 and R_2 to estimate the probability of an HU being touched by crime.

CHAPTER 4

MODEL-BASED ESTIMATORS

In this chapter we consider several superpopulation models of victimization and derive an estimate of θ on the basis of each. The fit of each model is tested using a χ^2 goodness of fit statistic for each of the years 1973 though 1975. We find that a correlated Bernoulli model is an inappropriate model of crime, while a homogeneous Bernoulli model and a Markov model provide better fits to the data. None of these models fits the data as well as the model under which the modified ad hoc estimator is consistent.

4.1. A Homogeneous Bernoulli Model

The first model-based estimator that we will discuss is based on a homogeneous Bernoulli model of victimization. For i = 1,...,H and j = 1,...,12 let

$$X_{ij} = \begin{cases} 1 \text{ if HU } i \text{ is victimized in month } j \\ \\ 0 \text{ otherwise .} \end{cases}$$

Note that this notation differs from the Y_{ij}'s in the previous chapter (Y_{ij} = 1 - X_{ij}). If we assume that the X_{ij}'s are independent Bernoulli random variables with a common value of $p = \Pr\{x_{ij}=1\}$, then the likelihood of each observed month j of data for HU i is

$$f(x_{ij} \mid p) = p^{x_{ij}} (1 - p)^{1-x_{ij}} \tag{4-1}$$

where p is unknown. This f is not to be confused with the $f(i,\theta)$ of Chapter 3.

Here f is the probability mass function of x_{ij} given p, whereas, in Chapter 3, $f(i,\theta)$ represented the probability of the specific outcome in which all of the i months of information that an HU contributed in a year were crime-free months. (For the homogeneous Bernoulli model, $f(i,\theta) = \theta^{i/12}$.)

Under this model, every HU has the same probability p of being victimized in any month. A given HU is victimized in month j independently of whether or not it is victimized in month l ($l \neq j$), and HU i is victimized independently of HU k ($k \neq i$).

The density of the complete data **X** is then

$$f(\mathbf{x}\,|\,p) = \prod_{i=1}^{H} \prod_{j=1}^{12} p^{x_{ij}} (1-p)^{1-x_{ij}} . \tag{4-2}$$

If we assume that the sampling and nonresponse mechanisms are ignorable then likelihood inferences for p can be based on the likelihood

$$L(p\,|\,o) = \sum_{0}^{1} \cdots \sum_{0}^{1} f(\mathbf{x}\,|\,p) \quad (x_{ij}^{0}:\ x_{ij}^{0} \text{ is unobserved})$$

$$= \prod_{((i,j):\ x_{ij} \text{ is observed})} p^{x_{ij}} (1-p)^{1-x_{ij}} \tag{4-3}$$

where o, as before, is the set of x_{ij}'s that were sampled and also responded. It is easy to show that the maximum likelihood estimate of p is

$$\hat{p} = \frac{\Sigma_{obs} x_{ij}}{T} , \tag{4-4}$$

where T is the total number of X_{ij}'s that are observed and

$$\Sigma_{obs} = \sum_{((i,j):\ x_{ij} \text{ is observed})} (\cdot) .$$

If we wish to estimate θ, the probability that an HU is crime-free for the year, then we need only note that

$$\theta = (1 - p)^{12} , \tag{4-5}$$

and thus the MLE of θ is

$$\hat{\theta}_3 = (1 - \hat{p})^{12} = \left(\frac{T - \Sigma_{obs} x_{ij}}{T}\right)^{12}$$

$$= \left(\frac{\text{number of crime-free months observed}}{T}\right)^{12} . \tag{4-6}$$

Table 9 shows the quantities needed to calculate $\hat{\theta}_2$ as well as the estimates for each of the years 1973-1975.

TABLE 9

Homogeneous Bernoulli Estimates

	1973	1974	1975
$\Sigma_{obs} x_{ij}$	363	372	327
T	8182	8461	8535
$1-\hat{p}$	.957	.956	.962
$.\hat{\theta}_3$	.580	.583	.626

Since the term $T - \Sigma_{obs} x_{ij}$ that appears in the numerator of $\hat{\theta}_3$ has a binomial distribution, we can use the formula given in Johnson and Kotz(1969) for the moments of a binomially distributed random variable to find the mean and variance of $\hat{\theta}_3$. Using this formula we find

$$E(\hat{\theta}_3) = \frac{1}{T^{12}} E(T - \Sigma_{obs} x_{ij})$$

$$= \frac{1}{T^{12}} \sum_{j=1}^{12} \binom{T}{j} (1-p)^j \left[\sum_{k=0}^{j} \binom{j}{k} (-1)^k (j-k)^{12} \right] \tag{4-7}$$

and

$$E(\hat{\theta}_3{}^2) = \frac{1}{T^{24}} \sum_{j=1}^{24} \binom{T}{j} (1-p)^j \left[\sum_{k=0}^{j} \binom{j}{k} (-1)^k (j-k)^{24} \right] . \tag{4-8}$$

We can then, of course, evaluate $Var(\hat{\theta}_3)$ as $E(\hat{\theta}_3{}^2) - [E(\hat{\theta}_3)]^2$. The actual numerical evaluation of expressions (4-7) and (4-8) requires some care because the term "T choose j" is very large for the larger values of j in the range of 1 to 24, and so a straightforward evaluation of either of these formulae would result in computer overflows. On the other hand, T^{12} and T^{24} are so large that division of each term by either of them results in computer underflow. In order to avoid these problems, in the evaluation of $E(\hat{\theta}_3)$ the bracketed part of (4-7) was first calculated for each value of j and then the expression

$$\log\left\{ \binom{T}{j} \frac{1}{T^{12}} (1-p)^j \left[\sum_{k=0}^{j} \binom{j}{k} (-1)^k (j-k)^{12} \right]\right\}$$

$$= \sum_{k=1}^{j} \log(T+1-k) - \sum_{k=1}^{j} \log k + \log \left[\sum_{k=0}^{j} \binom{j}{k} (-1)^k (j-k)^{12} \right]$$

$$+ j \log (1-p) - 12 \log T . \tag{4-9}$$

was exponentiated for each value of j and summed. The only underflow occurred when equation (4-9) was exponentiated with j=1 and for this value of j it was easy to verify that the contribution to equation (4-7) was negligible.

A similar procedure was used for evaluating expression (4-8). Again, several underflows occurred but it was easily seen that the contributions to $E(\hat{\theta}_3^2)$ for the small values of j were negligible. In fact, most of the contribution comes from the terms for j=22 through 24.

The results for the years 1973 through 1975 are presented in Table 10 along with the values of $\hat{\theta}_3$. The program that was used to compute the estimated standard deviations is program 2 in Appendix B.

Alternatively, we could use the δ method (Bishop et al., 1975) to find the asymptotic distribution of $\hat{\theta}_3$. If we let $q = 1 - p$ and $\hat{q} = 1 - \hat{p}$, then $\hat{\theta}_3 = \hat{q}^{12}$. Since $T^{1/2}(\hat{q} - q)$ is asymptotically normally distributed with mean 0 and variance pq, and since the derivative of $\hat{\theta}_3$ with respect to q is $12q^{11}$, we can use the δ method to find that $T^{1/2}(\hat{\theta}_3 - \theta)$ is asymptotically normally distributed with mean 0 and variance $144q^{23}p$. Hence for large T, $\hat{\theta}_3$ has an approximate normal distribution with variance $T^{-1}144q^{23}p$ and thus we can estimate the variance of $\hat{\theta}_3$ by $T^{-1}144\hat{q}^{23}\hat{p}$. The estimates of the standard deviation of $\hat{\theta}_3$ computed using this formula are .017, .016, and .016 respectively for the three years of data. These estimates are slightly larger than those shown in Table 10.

Both sets of estimated standard deviations are relatively small and what may concern us more than the variance is the fit of the homogeneous Bernoulli model. To get an idea of the appropriateness of the model, the HUs have been divided into 12 groups with the ith group containing all the HUs that contributed exactly i months of information during the year of interest. The observed and estimated expected number of victimized HUs were then calculated for each of the 12 groups. Under the Bernoulli model, the probability that an HU that contributes i months of information does not report a victimization is p^i and so the probability that the HU is victimized during those i months is $1 - p^i$.

TABLE 10

Estimated Moments and Standard Deviations of $\hat{\theta}_3$

	$\hat{\theta}_3$	est. $E(\hat{\theta}_3)$	est. $E(\hat{\theta}_3^2)$	est. s.d. $\hat{\theta}_3$
1973	.5801	.5804	.3370	.012
1974	.5830	.5834	.3404	.011
1975	.6258	.6261	.3921	.010

The number of HUs that contributed i months of information, the observed and estimated expected number of victimized HUs, and the contribution to a χ^2 statistic for i = 1, ..., 12 and for each of the years 1973 through 1975 are displayed in Table 11.

The corresponding χ^2 goodness of fit statistics for the three years are 35.82, 31.59, and 15.53, respectively, and each has 10 degrees of freedom. Thus, at least for 1973 and 1974, the fit is relatively poor. Note that the fit is particularly bad for i = 12 in each year and that for these HUs the number of victimized HUs has been overestimated. Since HUs that contribute 12 months of information cannot have had any nonresponse in the year of interest, this may be an indication that the nonresponse data are not missing at random and that HUs with missing data tend to be more highly victimized than HUs with full data.

TABLE 11

Fit of Homogeneous Bernoulli Model

	i	HUs contributing i months	obs. # vict.	est. exp. # vict.	$(O-E)^2/E$
1973	1	24	4	1.066	8.081
	2	31	4	2.691	0.637
	3	24	3	3.058	0.001
	4	25	5	4.152	0.173
	5	25	6	5.077	0.168
	6	71	14	16.933	0.508
	7	21	6	5.718	0.014
	8	12	2	3.655	0.749
	9	21	10	7.045	1.239
	10	11	3	4.015	0.257
	11	21	13	8.257	2.724
	12	550	161	231.11	21.268
					35.82
1974	1	40	1	1.760	0.328
	2	46	5	3.961	0.273
	3	42	9	5.305	2.574
	4	38	3	6.259	1.697
	5	49	15	9.874	2.662
	6	94	23	22.241	0.026
	7	48	13	12.970	0.000
	8	40	15	12.092	0.699
	9	41	7	13.653	3.242
	10	42	18	15.221	0.507
	11	35	10	13.66	0.982
	12	451	129	188.157	18.599
					31.59
1975	1	30	2	1.149	0.630
	2	42	2	3.154	0.422
	3	39	5	4.313	0.109
	4	48	5	6.941	0.543
	5	41	8	7.273	0.073
	6	89	20	18.592	0.107
	7	31	8	7.415	0.046
	8	50	9	13.415	1.453
	9	41	6	12.152	3.115
	10	49	18	15.842	0.294
	11	43	12	15.016	0.606
	12	452	132	169.093	8.137
					15.53

4.2. A Correlated Bernoulli Model

In this section we discuss a model that weakens the assumption of independence from month to month for each HU by allowing for correlation among the observations for a particular HU. The following distribution is adapted from Tallis(1962). For $i = 1,\dots,H$, $0 \le p \le 1$, and $0 \le \rho \le 1$, let the probability density for the observations from HU i be

$$f_i(x_{i1},\dots,x_{i12} \mid p,\rho) = \begin{cases} q^{12}(1-\rho) + \rho q & \text{if } x_{i1}=x_{i2}=\dots=x_{i12}=0 \\ p^{12}(1-\rho) + \rho p & \text{if } x_{i1}=x_{i2}=\dots=x_{i12}=1 \\ q^{12-x_{i+}}\, p^{x_{i+}}\,(1-\rho) & \text{if not all the } x_{ij}\text{'s are equal} \end{cases} \tag{4-10}$$

where $q = 1 - p$ and $x_{i+} = \sum_{j=1}^{12} x_{ij}$. It follows that X_{ij} has a Bernoulli distribution with parameter p ($i = 1,\dots,H$; $j = 1,\dots,12$) and that the correlation between X_{ij} and X_{il} is ρ ($j \ne l$, $i = 1,\dots,H$). We assume that $(X_{i1},\dots,X_{i12})$ is independent of $(X_{k1},\dots,X_{k12})$ for $i \ne k$.

As with the previous model, every HU has the same probability p of being victimized in any month, and HU i is victimized independently of HU k ($k \ne i$). This model has the feature that, when $0 < \rho \le 1$, X_{ij} is no longer independent of X_{il}. The correlation between any two, not necessarily consecutive, months of data is ρ. Note that when $\rho = 0$, this model simplifies to the Bernoulli model with complete independence of monthly observations discussed in the previous section. The fact that the homogeneous Bernoulli model overestimated the number of HUs that were victimized among those contributing 12 months of information leads us to believe that victimization does not occur independently from month to month but rather that victimization in one month may increase the probability of victimization in other months. This model may allow us to take this correlation into account.

The density function of the complete data is then

$$f(\mathbf{X}|p,\rho) = \prod_{i=1}^{H} f_i(x_{i1},\ldots,x_{i12}|p,\rho)$$

$$= \prod_{i=1}^{H} \Big\{ [q^{12}(1-\rho) + \rho q]^{I(x_{i+}=0)} \times [p^{12}(1-\rho) + \rho p]^{I(x_{i+}=12)}$$

$$\times [p^{x_i} q^{12-x_{i+}} (1-\rho)]^{I(0<x_{i+}<12)} \Big\} . \qquad (4\text{-}11)$$

If we assume that the sampling and nonresponse mechanisms are ignorable, likelihood inferences can be based on

$$L(p,\rho|o) = \sum_{(x^0_{ij}:\, x^0_{ij} \text{ is unobserved})}^{1} \cdots \sum^{1} f(\mathbf{x}|p,\rho) \qquad (4\text{-}12)$$

where o is defined as before. If we define m_i to be the number of months for which the data for HU i is missing (either out of sample or nonresponse) and v_i to be the number of months in which HU i was victimized ($v_i = \sum_{j:\, x_{ij} \text{ is observed}} x_{ij}$), then expression (4-12) can be shown to be equal to

$$\prod_{i:m_i<12} \Big\{ [q^{12}(1-\rho) + \rho q]^{I(v_i=0)} [p^{12}(1-\rho) + \rho p]^{I(v_i=12)}$$

$$\times [p^{v_i} q^{12-v_i} (1-\rho)]^{I(0<v_i<12)}$$

$$+ I(m_i>1) \sum_{j=1}^{m_i-1} \Big[\binom{m_i}{j} p^{v_i+j} q^{12-j-v_i} \Big] (1-\rho)$$

$$+ I(m_i>0)\, [p^{12}(1-\rho) + \rho p]^{I(v_i=12-m_i)} [p^{v_i+m_i} q^{12-v_i-m_i} (1-\rho)]^{I(v_i<12-m_i)} \Big\} . \qquad (4\text{-}13)$$

Maximum likelihood estimates of p and ρ can be obtained by iterative procedures, and the maximum likelihood estimate of θ, the probability that an HU is crime-free for a year, is then

$$\hat{\theta}_4 = (1-\hat{\rho})(1-\hat{p})^{12} + \hat{\rho}(1-\hat{p}) . \qquad (4\text{-}14)$$

Note that $\hat{\theta}_4$ is a linear combination of $\hat{\theta}_3$, the Bernoulli estimator, and $1-\hat{p}$, the estimator that results from perfect correlation, i.e. once crime-free, always crime-free, once victimized, always victimized. The density (4-10) is itself a mixture of the binomial density and the density that assigns probability p to the outcome $x_{i1} = \dots = x_{i12} = 0$ and probability q to the outcome $x_{i1} = \dots = x_{i12} = 1$. The mixing probabilities are $1-\rho$ and ρ respectively.

The logarithm of the likelihood given in equation (4-13) has been maximized using the IMSL subroutine ZXMIN which uses a quasi-Newton method. (Actually, -1 times the log likelihood was minimized.) In order to force the subroutine to find a maximum that was in the parameter space, $0 \le p \le 1$ and $0 \le \rho \le 1$, a penalty function was used. For any value outside of the parameter space, -1 times the log likelihood was set equal to a large number. The values of p and ρ resulting from this maximization are displayed in Table 12 along with the corresponding values of θ. The program that was used to maximize the likelihood is program 3 in Appendix B.

TABLE 12

Correlated Bernoulli Estimates

	1973	1974	1975
$\hat{p}$	.072	.066	.052
$\hat{\rho}$	.417	.360	.297
$\hat{\theta}_4$	.626	.619	.651

In order to verify that the maximum that was obtained by ZXMIN is the global maximum, the iterative procedure was started from several different initial points. Regardless of the starting point, the routine always converged to the same values. In addition, the log likelihood surface given the 1973 data has been plotted in Figures 5 and 6. The plots were generated using

DISSPLA[2], a library of plotting subroutines. We can see from the two views of the log likelihood surface that the function is uni-modal and thus the maximum computed by ZXMIN is a global maximum. Near the maximum the surface is much more peaked with respect to the variable p than with respect to ρ. Thus the variance of $\hat{\rho}$ will be large compared to the variance of $\hat{p}$. Again, as in the discussion of Figures 3 and 4, we need to keep in mind that these plots are plots of the log likelihood and that the likelihood itself will be much more peaked than the log likelihood.

As we will see, there is no need to worry about the size of the variances for this model since the fit of the model is so poor! Table 13 shows the values of the density, $f(x_1, \ldots, x_{12} | p, \rho)$, for $p = .072$ and $\rho = .417$. Notice that this density has a spike at $\Sigma x_j = 12$. According to this density, it is more likely that an HU will be victimized at least once in every month than that the HU will be victimized in exactly 1 month. This is a result of the previously noted fact that the density (4-10) is actually a mixture of 2 densities, one of which puts probability p on 12 crime-free months and probability q on 12 victimized months. In our 1% sample there are no HUs that have had 12 months of victimization and thus, for this perfect correlation model, we would expect the value of q to be very small. On the other hand, the second model in the mixture is a binomial model in which q is the probability of victimization in any particular month. For this model, we would expect a larger value of q. The result is that too much probability is put on the outcome $x_{i1} = \ldots = x_{i12} = 1$. The third column of Table 12 shows the actual number of HUs in 1973 that contributed 12 months of victimization information and reported the appropriate number of victimization months. The contributions to a χ^2 statistic when $\Sigma x_j = 1, 2, 3,$ or 4 are extremely large

[2]DISSPLA is distributed by Integrated Software Systems Corporation, 4186 Sorrento Valley Blvd., San Diego, CA 92121

FIGURE 5

The log likelihood of p and ρ
under the correlated Bernoulli model.
1973 data

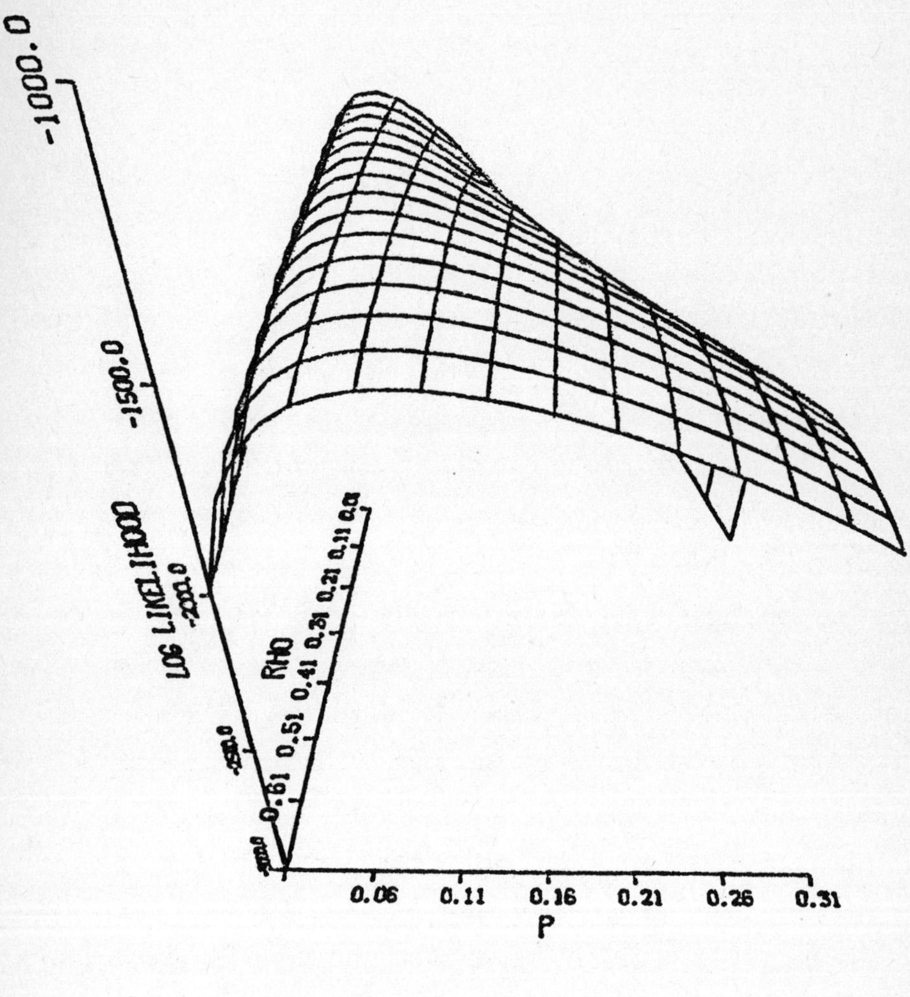

FIGURE 6

The log likelihood of p and ρ
under the correlated Bernoulli model.
1973 data

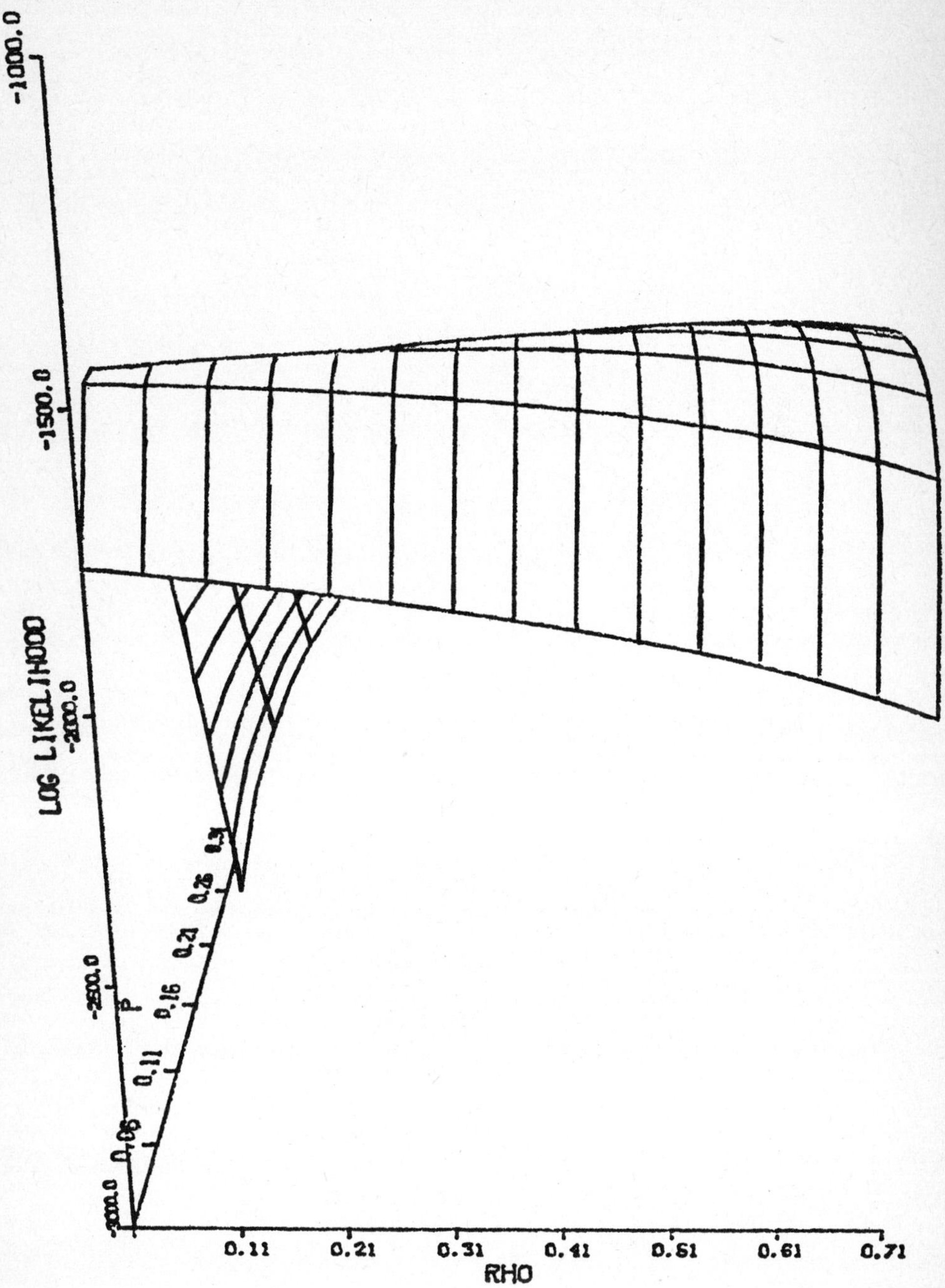

because the expected number of HUs (550 times the density) are very small relative to the observed number. The last row of Table 13 also contributes a large amount to the χ^2 statistic since we expect about 16 HUs and observe none in that row. Even when we combine the cells for Σx_j between 2 and 11, we still find a total χ^2 value of over 6000 with 1 degree of freedom! Clearly this model does not accurately describe the victimization phenomenon and it will not be discussed further.

TABLE 13

Correlated Bernoulli Density and Observations

Σx_j	density	observed # HUs
0	.6248	389
1	.0185	91
2	.0014	49
3	.0001	14
4	.0000	7
5-11	.0000	0
12	.0280	0

4.3. A Markov Model

In this section we discuss a Markov model of victimization. Let $X_{i1},\ldots,X_{i12}$ be defined as before. We can consider these observations as arising from a two state Markov chain with states 0 and 1, where 0 indicates that no victimization occurred during the month and 1 indicates that at least 1 victimization occurred. Let (p_0,p_1) be the initial probability vector and

$$\begin{pmatrix} p_{00} & p_{01} \\ p_{10} & p_{11} \end{pmatrix}$$

the transition matrix. Thus the probability that an HU is crime-free in January is p_0 and the probability that the HU moves from the crime-free state to the victimized state, given that it is in the crime-free state, is p_{01}, etc. This model allows victimization in a particular month, rather than being independent of victimization in the other months as in the homogeneous Bernoulli model, to depend on victimization in the previous month.

The likelihood of the complete data for HU i is then

$$L(p_0, p_{00}, p_{10} \mid x_{i1}, \ldots, x_{i12}) \propto p_0^{I(x_{i1}=0)} \, p_1^{I(x_{i1}=1)} \, p_{00}^{n_{00}^{(i)}} \, p_{01}^{n_{01}^{(i)}} \, p_{10}^{n_{10}^{(i)}} \, p_{11}^{n_{11}^{(i)}} \tag{4-15}$$

where $n_{jk}^{(i)}$ is the number of transitions from state j to state k for HU i. If we assume that HUs are victimized independently of each other then the likelihood of the complete data for the entire sample is

$$L(p_0, p_{00}, p_{10} \mid x) \propto p_0^{n_0} \, p_1^{n_1} \, p_{00}^{n_{00}} \, p_{01}^{n_{01}} \, p_{10}^{n_{10}} \, p_{11}^{n_{11}} \tag{4-16}$$

where $n_j = \sum_{i=1}^{H} I(x_{i1}=j)$, $j = 0, 1$; and $n_{jk} = \sum_{i=1}^{H} n_{jk}^{(i)}$.

If we assume that the sampling and nonresponse mechanisms are ignorable, likelihood inferences for the incomplete data can be based on

$$L(p_0, p_{00}, p_{10} \mid o) \propto \underset{(x_{ij}:\, x_{ij} \text{ is unobserved})}{\sum_0^1 \cdots \sum_0^1} L(p_0, p_{00}, p_{10} \mid x) . \tag{4-17}$$

Expression (4-17) can be evaluated numerically and maximum likelihood estimates of p_0, p_{00}, and p_{10} can be obtained by iterative procedures. The maximum likelihood estimate of θ, the proportion of crime-free HU's, is then

$$\hat{\theta}_5 = \hat{p}_0 (\hat{p}_{00})^{11} . \tag{4-18}$$

TABLE 14

MLEs and Estimated Standard Deviations

	1973	1974	1975
$\hat{p}_0$	.960	.953	.953
est. s.d. $\hat{p}_0$	$.707\times10^{-2}$	$.703\times10^{-2}$	$.708\times10^{-2}$
$\hat{p}_{00}$	.959	.959	.967
est. s.d. $\hat{p}_{00}$	$.235\times10^{-2}$	$.232\times10^{-2}$	$.207\times10^{-2}$
$\hat{p}_{10}$	.865	.882	.838
est. s.d. $\hat{p}_{10}$	$.252\times10^{-1}$	$.229\times10^{-3}$	$.377\times10^{-1}$
$\hat{\theta}_5$	.608	.604	.662
est. s.d. $\hat{\theta}_5$	$.170\times10^{-1}$	$.168\times10^{-1}$	$.164\times10^{-1}$

As with the correlated Bernoulli model, ZXMIN was used to find the MLEs which are given in Table 14. ZXMIN requires a subroutine that evaluates the likelihood (actually, -1 times the log likelihood in this case) for any values of the parameters. For the correlated Bernoulli model, we were able to simplify the expression for the likelihood of the parameters given only the observed data and thus calculation of that likelihood for any values of the parameters was easy. In the case of the Markov model the likelihood given in equation (4-17) is rather difficult to compute. In order to see how this likelihood can be computed, we first examine the likelihood given one HU that contributes 12 months of information:

$$p_0^{I(x_1=0)}\, p_1^{I(x_1=1)} \times p_{00}^{\sum_{j=1}^{11} I(x_j=0 \text{ and } x_{j+1}=0)} \times p_{10}^{\sum_{j=1}^{11} I(x_j=1 \text{ and } x_{j+1}=0)}$$

$$\times\, p_{01}^{\sum_{j=1}^{11} I(x_j=0 \text{ and } x_{j+1}=1)} \times p_{11}^{\sum_{j=1}^{11} I(x_j=1 \text{ and } x_{j+1}=1)} \,. \tag{4-19}$$

When any of the data are missing, we must sum the missing x_j's out of expression (4-19). There are four special cases that we need to consider:

1. the missing data occur at the end of the year -- this is the easiest case. When only the first k months of information are available, then summing expression (4-19) over the last 12 - k of the x_j's results in an expression of the same form but with the 11's replaced by k-1's.

2. the missing data occur at the beginning of the year. In this case we must add the probabilities of all the possible observations that end with the observed data. These possible observations are generated by the computer program, the probabilities are computed and summed.

3. the missing data occur in the middle of the year. Note that because of the survey design there must be exactly 6 months of missing data in this case. As in the previous case, we must add the probabilities of all of the possible observations that begin with and end with our observed data. The $2^6 = 64$ possible sequences of missing data are generated by the computer and their probabilities are calculated and summed.

4. the missing data occur at the beginning and at the end of the year. This is really just a combination of the first two cases and is handled similarly.

The contribution of an HU to the likelihood takes one of these four forms. The logarithm of each contribution is calculated, multiplied by the number of corresponding HUs, and added to the likelihood. The program that actually does the computation and maximization is program 4 in Appendix B. As before, a penalty function was used to assure that the procedure searched for a maximum that was in the parameter space.

In order to determine if the log likelihood was unimodal, the function was plotted. Since the likelihood is a function of the three parameters p_0, p_{00}, and p_{10} it was not possible to plot it directly. Thus it was first plotted as a function of p_0 and p_{00} with p_{10} set equal to its 1975 MLE as found by ZXMIN. Next p_{00} was set equal to its 1975 MLE and the log likelihood was plotted as a function of p_0 and p_{10}, and similarly for p_0. These plots were produced by DISSPLA and are shown in Figures 7 through 11. From them, it appears as though the likelihood is unimodal and fairly smooth. By close examination, we can see that the values obtained by ZXMIN do correspond to maxima in the plots. Some of the plots seem very flat but once again we must remember that we are looking at the logarithm of the likelihood over large parts of the parameter space.

In Figures 7 and 8 p_{10} is set equal to .838. We see that the surface is much more curved in the p_{00} direction than in the p_0 direction. This is what we would have expected since there are many more 0 to 0 transitions than initial observations of the state 0. In Figure 9 p_{00} is fixed at .967. We see that this log likelihood surface is extremely flat. Thus the variances of $\hat{p}_0$ and $\hat{p}_{10}$ will be large relative to the variance of $\hat{p}_{00}$. In Figures 10 and 11 $\hat{p}_0$ is set to .954. This surface is curved with respect to p_{00} much more than with respect to p_{10} and so the variance of $\hat{p}_{00}$ will be small relative to the

FIGURE 7

The log likelihood, under the Markov model, of p_0 and p_{00} with p_{10} = .838.

1975 data

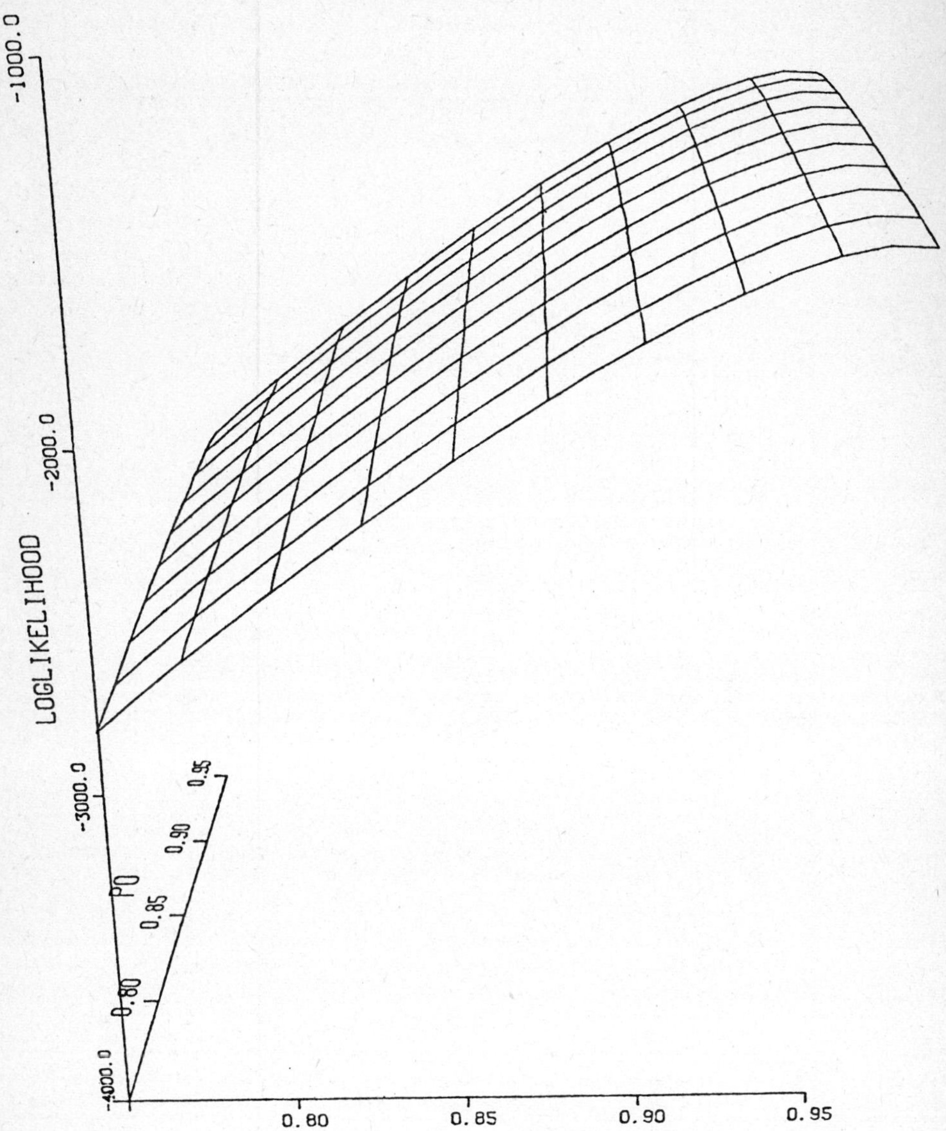

FIGURE 8

The log likelihood, under the Markov model, of p_0 and p_{00} with p_{10} = .838.

1975 data

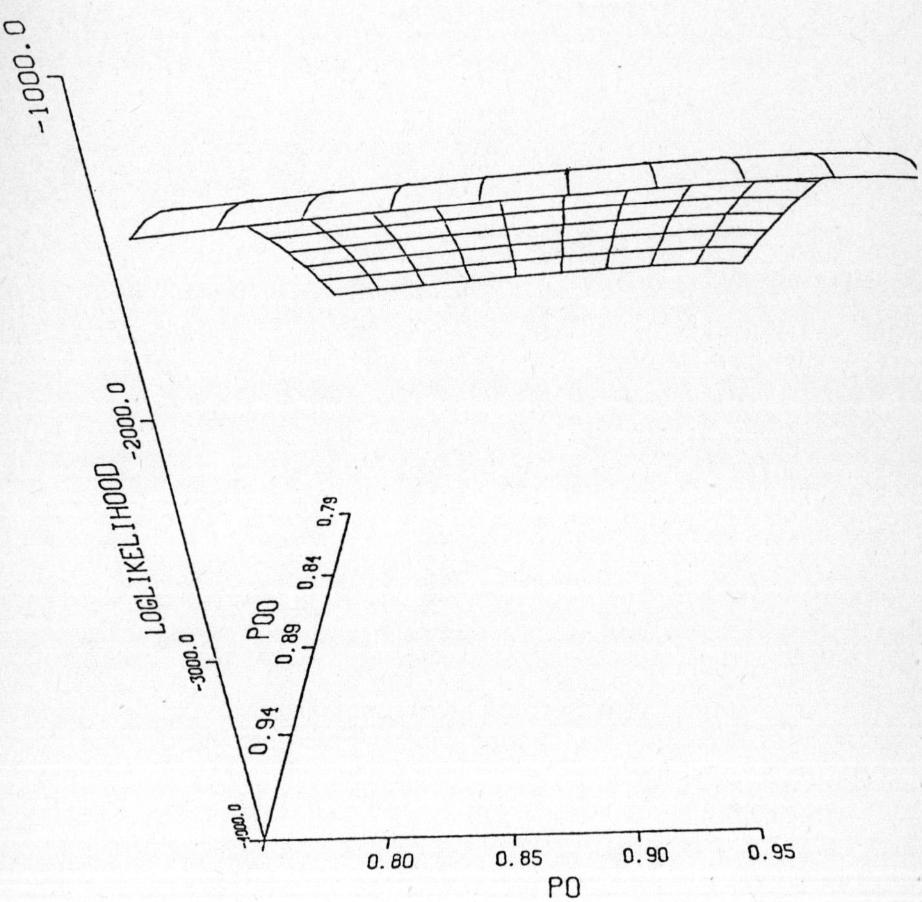

FIGURE 9

The log likelihood, under the Markov model, of p_0 and p_{10} with p_{00} = .967.

1975 data

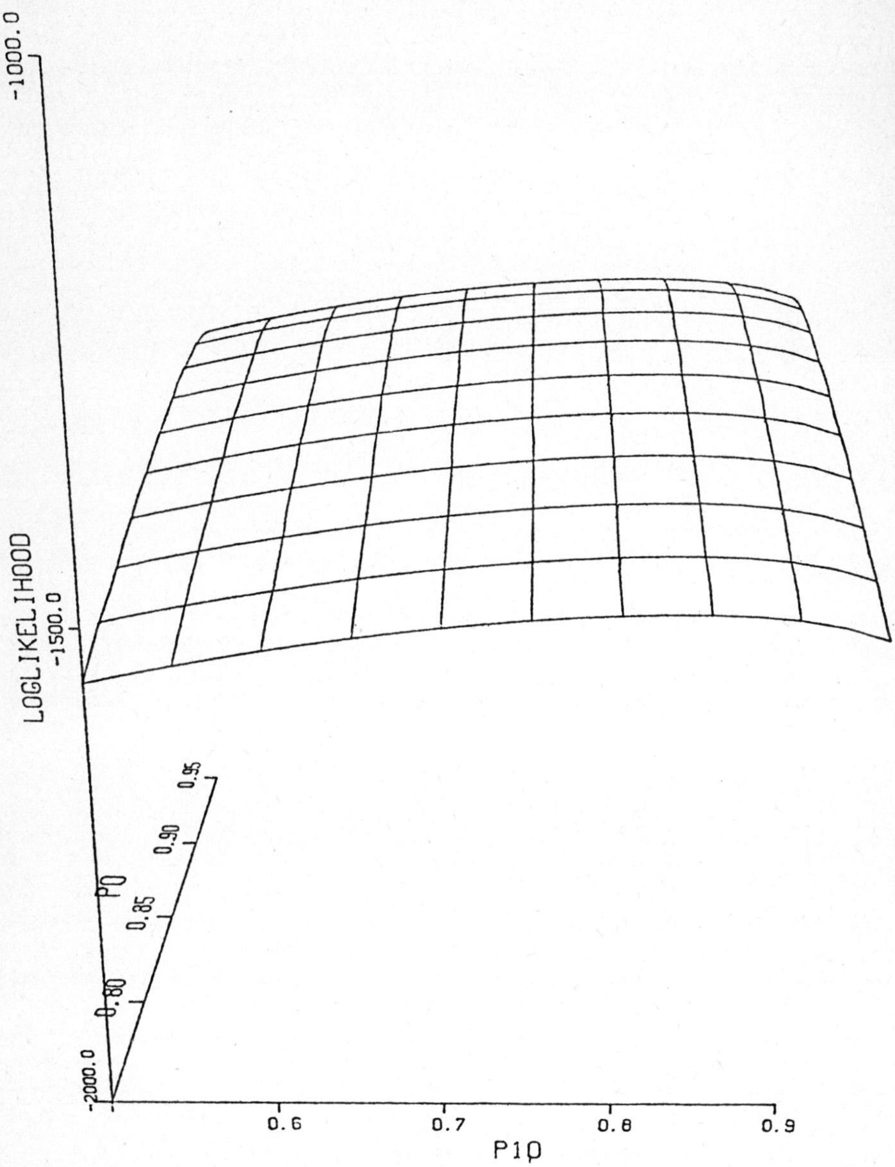

FIGURE 10

The log likelihood, under the Markov model, of p_{00} and p_{10} with p_0 = .954.

1975 data

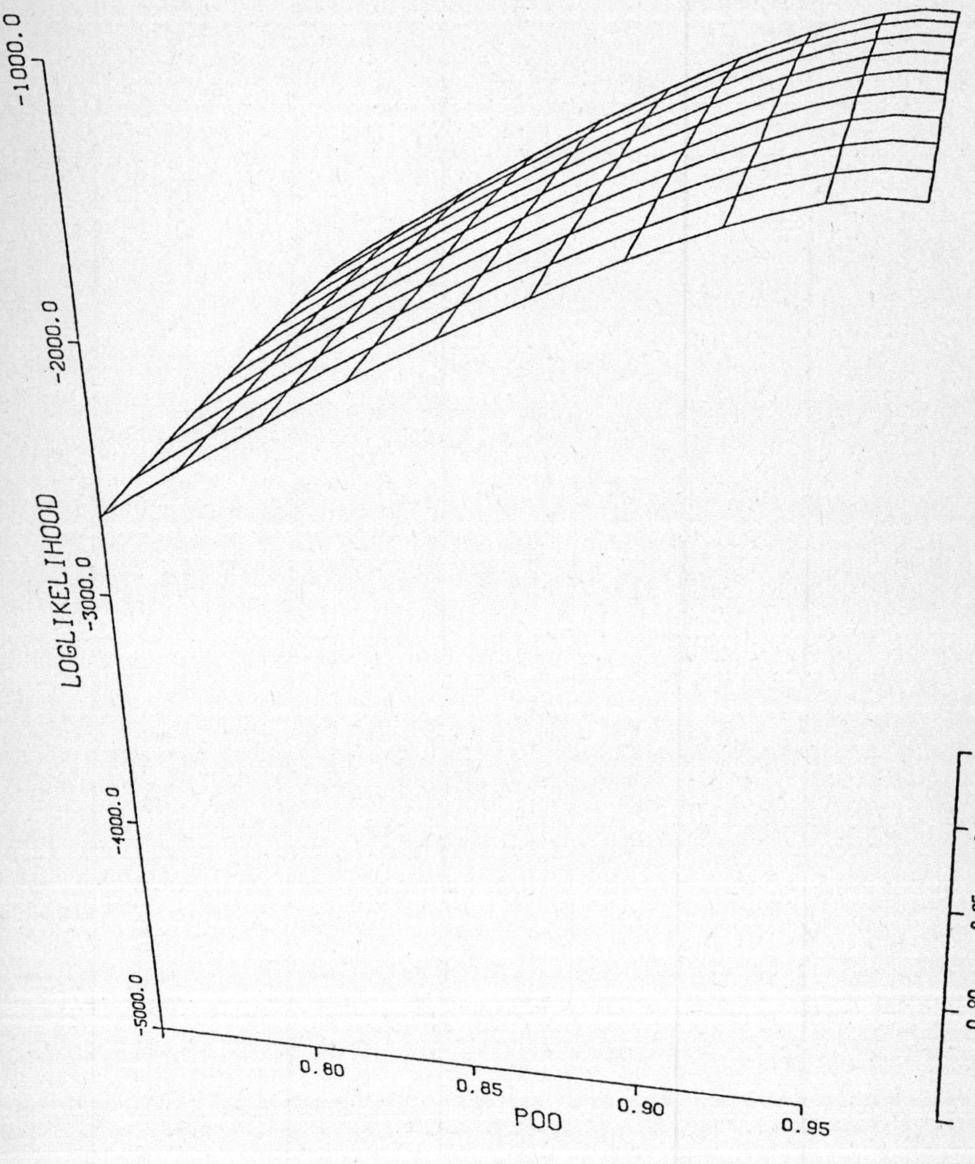

FIGURE 11

The log likelihood, under the Markov model, of p_{00} and p_{10} with p_0 = .954.

1975 data

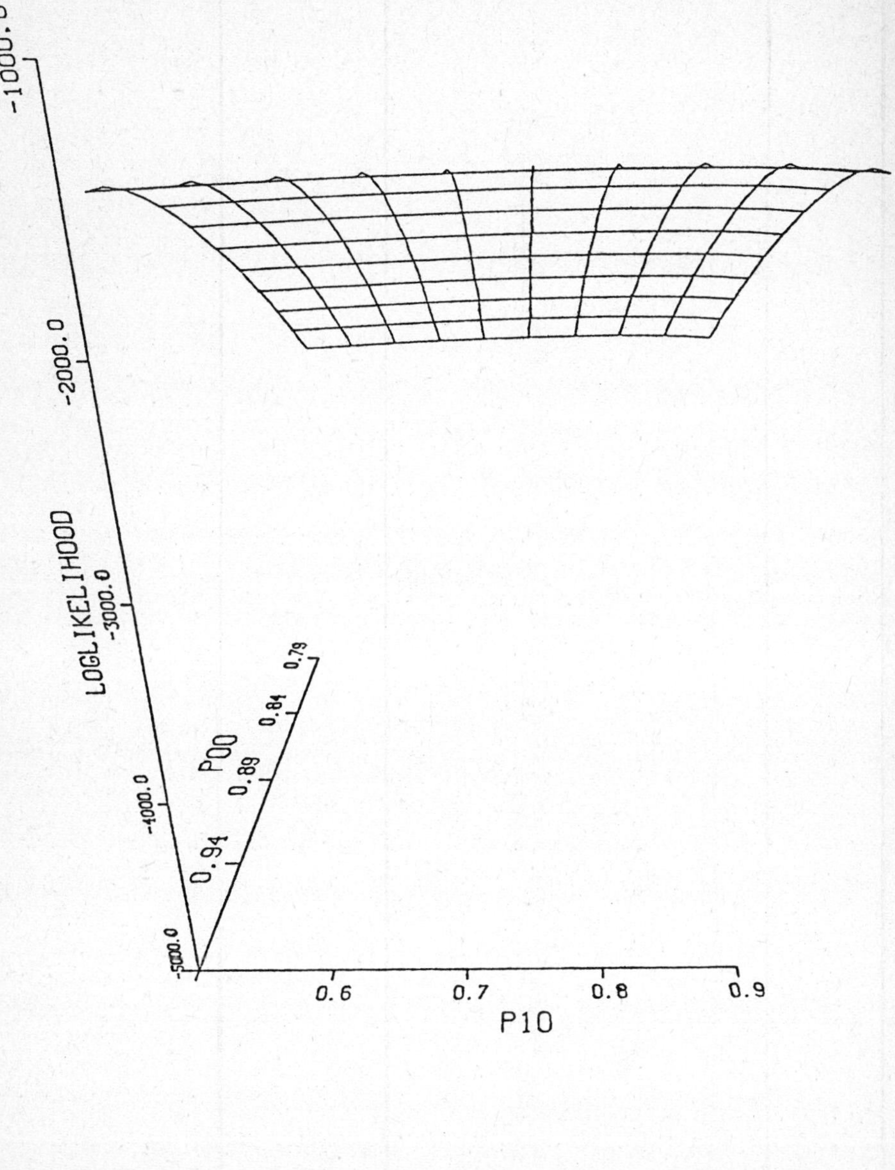

variance of $\hat{p}_{10}$. This again is as we would have expected since we have many more observations in the state 0 than in the state 1, and hence more information about 0 to 0 transitions than about 1 to 0 transitions.

We can estimate the variances of the MLEs using asymptotic methods. In order to do this, we first need an estimate of the Fisher information which involves evaluating the matrix of second derivatives of the log likelihood. As discussed above, the log likelihood consists of three distinctly different pieces (since the fourth can be considered as a combination of the first two). Thus the second derivatives must be calculated for each of these three different parts. Since there are three first derivatives and six second derivatives to compute for each of the three parts of the log likelihood, the computations are fairly long. The program that was written to perform these computations is program 5 of Appendix B.

Once the estimated information matrix for (p_0, p_{00}, p_{10}) was obtained, it was necessary to invert the information matrix and then to transform this estimate of the variance-covariance matrix to one from which a variance estimate for $\hat{\theta}_5$ could be found. The transformation that was chosen was from (p_0, p_{00}, p_{10}) to $(p_0, \theta = p_0 p_{00}^{11}, p_{10})$. The matrix of first derivatives for this transformation is

$$A = \begin{pmatrix} 1 & p_{00}^{11} & 0 \\ 0 & 11 p_0 p_{00}^{10} & 0 \\ 0 & 0 & 1 \end{pmatrix} \tag{4-20}$$

and thus (cf. Cox and Hinkley, 1974) the estimated variance-covariance matrix for (p_0, θ, p_{10}) is

$$V(p_0, \theta, p_{10}) = \mathbf{A}\, I^{-1}(p_0, p_{00}, p_{10})\, \mathbf{A}^t . \tag{4-21}$$

The 2,2 element of $V(p_0,\ \theta,\ p_{10})$ is then an estimate of the variance of θ. The estimated information matrix and its inverse, as well as V, for each of the three years 1973-1975 are shown in Table 15.

TABLE 15

Estimated Information and Variances for the Markov Model

Year	$I(p_0,p_{00},p_{10})$			$I^{-1}(p_0,p_{00},p_{10})$		
1973	$.200\text{x}10^{5}$			$.500\text{x}10^{-4}$		
	$.558\text{x}10^{2}$	$.181\text{x}10^{6}$		$-.156\text{x}10^{-7}$	$.552\text{x}10^{-5}$	
	$-.684\text{x}10^{1}$	$.114\text{x}10^{3}$	$.157\text{x}10^{4}$	$.219\text{x}10^{-6}$	$-.400\text{x}10^{-6}$	$.637\text{x}10^{-3}$
1974	$.203\text{x}10^{5}$			$.494\text{x}10^{-4}$		
	$.610\text{x}10^{2}$	$.185\text{x}10^{6}$		$-.165\text{x}10^{-7}$	$.540\text{x}10^{-5}$	
	$-.175\text{x}10^{2}$	$.119\text{x}10^{3}$	$.190\text{x}10^{4}$	$.454\text{x}10^{-6}$	$-.338\text{x}10^{-6}$	$.526\text{x}10^{-3}$
1975	$.199\text{x}10^{5}$			$.512\text{x}10^{-4}$		
	$.122\text{x}10^{3}$	$.234\text{x}10^{6}$		$-.279\text{x}10^{-7}$	$.132\text{x}10^{-5}$	
	$-.185\text{x}10^{2}$	$.141\text{x}10^{3}$	$.704\text{x}10^{3}$	$.132\text{x}10^{-5}$	$-.860\text{x}10^{-6}$	$.142\text{x}10^{-2}$

Year	$V(p_0,\theta,p_{10})$		
1973	$.540\text{x}10^{-4}$		
	$.341\text{x}10^{-4}$	$.290\text{x}10^{-3}$	
	$.236\text{x}10^{-6}$	$-.264\text{x}10^{-5}$	$.637\text{x}10^{-3}$
1974	$.534\text{x}10^{-4}$		
	$.338\text{x}10^{-4}$	$.281\text{x}10^{-3}$	
	$.492\text{x}10^{-6}$	$-.203\text{x}10^{-5}$	$.526\text{x}10^{-3}$
1975	$.557\text{x}10^{-4}$		
	$.385\text{x}10^{-4}$	$.269\text{x}10^{-3}$	
	$.147\text{x}10^{-5}$	$-.542\text{x}10^{-5}$	$.142\text{x}10^{-1}$

Using these estimated variance-covariance matrices we have computed the estimated standard deviations for the MLEs and included them along with the MLEs, in Table 14. It appears that the 1974 Markov estimate is slightly lower than the 1973 estimate while the 1975 estimate is much higher than the 1974 estimate.

In all of the preceding calculations, double precision variables were used. ZXMIN was able to find the MLEs to about 8 digits of accuracy. The matrix inversions were done by the double precision IMSL subroutine LINV2F. The inversion of the estimated information matrices for (p_0, θ, p_{10}) was done in two ways. First, $V(p_0, \theta, p_{10})$ was calculated using formula (4-21) and inverted. As a second method, $I(p_0, p_{00}, p_{10})$ was computed and then $V(p_0, \theta, p_{10})$ was calculated using :

$$V(p_0, \theta, p_{10}) = [(A^{-1})^t I(p_0, p_{00}, p_{10})\, A^{-1}]^{-1} . \tag{4-22}$$

It is easy to verify that

$$A^{-1} = \begin{pmatrix} 1 & -p_{00}/(11p_0) & 0 \\ 0 & 1/(11p_0 p_{00}{}^{10}) & 0 \\ 0 & 0 & 1 \end{pmatrix},$$

and thus expression (4-22) requires only one matrix inversion. The results from the two different methods of inversion were virtually the same and so only the results of the first method are included.

To test the fit of the Markov model, the HUs were separated into 29 different groups, each having a unique probability under the Markov model of being victimized during the given year. The 29 groups are as follows:

1-12 i months of data followed by 12 - i months missing, i = 1, ..., 12

13 6 months missing in the middle -- regardless of which 6 months are missing, the probability of victimization is the same under this model

14-24 12 - i months of missing data followed by i months of data, i = 1, ..., 11

25-29 i months of missing data followed by 6 months of information, followed by 6 - i months of missing data, i = 1, ..., 5

The observed and expected numbers of victimized HUs that fell into each of these categories were computed from the 1973 data. As with evaluating the likelihood and the information matrix, it was necessary to write a computer program to compute the expected numbers of victimized HUs. The results are displayed in Table 16.

Again we see that this model has overestimated by a fairly large amount the number of victimized HUs among the HUs that have contributed 12 months of information. This may be an indication that HUs with nonresponse and possibly HUs that have just been rotated in or out of the survey have different victimization rates from those that are in the survey for the full 12 months.

In order to make comparisons with some of the previous fits, and to reduce the sparsity of Table 16, the 29 groups have been combined into 12 with all the HUs in each group having contributed the same number of months of information. These results are shown in Table 17.

Again we find that the fit is poor, especially for i=1 and i=12. In Table 11 we saw that the fit of the homogeneous Bernoulli model was poor for i=1 in 1973, but not for the other 2 years. This is true for the Markov model also. In 1974, one of the 40 HUs that contributed exactly one month of data was victimized while the estimated expected number of victimized HUs was 1.823. For 1975, the estimated number was 1.166 and the observed number was 2. The contributions to a χ^2 statistic were .371 and .597, respectively, which are much less than the corresponding 7.959 for 1973. For i=12, the fit of the 1974

TABLE 16

Observed and Expected Counts for the Markov Model

i	#HUs in i months, missing 12-i	# vict. HUs	exp. # vict. HUs
1	1	0	.040
2	5	2	.393
3	6	2	.697
4	3	1	.457
5	4	2	.747
6	18	2	3.959
7	12	3	3.021
8	7	2	1.976
9	12	5	3.738
10	6	2	2.038
11	3	8	4.765
12	550	161	215.781

i	#HUs in i, missing 6, in 6-i	#vict. HUs	exp. # vict. HUs
1-5	15	2	3.351

i	# HUs missing 12-i months, in i	#vict. HUs	exp # vict. HUs
1	23	4	1.035
2	26	2	2.182
3	18	1	2.183
4	22	4	3.456
5	21	4	4.021
6	23	6	5.162
7	9	3	2.304
8	5	0	1.432
9	9	5	2.839
10	5	1	1.717
11	8	5	2.958

i	#HUs missing i, in 6, missing 6-i	# vict. HUs	exp. # vict. HUs
1	4	0	.896
2	5	1	1.122
3	1	0	.224
4	2	1	.449
5	3	2	.683

TABLE 17

Fit of Markov Model
1973 data

i	#HUs in i months	# vict. HUs	exp. # vict. HUs	$(O-E)^2/E$
1	24	4	1.075	7.959
2	31	4	2.575	.789
3	24	3	2.880	.005
4	25	5	3.913	.302
5	25	6	4.768	.318
6	71	14	15.836	.213
7	21	6	5.325	.086
8	12	3	3.408	.152
9	21	10	6.577	1.781
10	11	3	3.755	.152
11	21	13	7.723	3.606
12	550	161	215.781	13.907
				29.699

data is still poor - 178.596 expected and 129 observed for a χ^2 contribution of 13.773. For 1975, the fit at i=12 is much better with 152.776 estimated expected victimized HUs and 132 observed victimized HUs for a χ^2 contribution of 2.82. This, however, is still larger than the contribution from any other value of i.

The overall χ^2 values are 29.699, 27.970, and 9.712 respectively, each with 8 degrees of freedom. Thus, as is the case with the homogeneous Bernoulli model, the fit is rather poor for 1973 and 1974, but seems to be fairly good for 1975. As with the previous models, the Markov model overestimates the number of HUs contributing 12 months of information that are victimized for each of the three years. This may indicate that the missing data are not missing at random. We examine this possibility in more detail in Chapter 5.

4.4. Comparisons of Estimates

In Chapters 3 and 4, seven different estimators of the cheery indicator of crime have been discussed. The estimates, which have been computed from the one percent sample of the data, are shown in Figure 12.

In Chapter 3, the values of $\hat{\theta}_2$, the MLEs of the model for which the modified ad hoc estimator is consistent, were computed using only data from HUs that had not had any nonresponse. Here they have been computed using all of the data in order to be able to make comparisons. The estimates themselves are .672, .667, and .695 with corresponding estimated standard deviations .017, .009, and .007. The associated χ^2 statistics are 6.17, 6.20, and 2.00 with 10 degrees of freedom and so we see that the fit is still good. Of the 4 models discussed in Chapters 3 and 4, this one seems to fit the data best. From Figure 12 we see that the MLEs for this model are very close to the modified ad hoc estimates and these estimates are in agreement with the BJS estimates. We notice, rather surprisingly, that the cheery indicators appear to increase over time indicating a decreasing prevalence of crime. The increase in the consistent estimate from 1973 to 1975 is .023 with an estimated standard deviation of .027 and thus is not a significant increase. The increase from 1974 to 1975 is .028 with an estimated standard deviation of .012 and thus this increase is at best marginally significant. Even though we do not have sufficient evidence to suggest that the prevalence of crime ($1 - \theta$) is falling, we see that, according to this sample of NCS data, it is not rising.

FIGURE 12

Estimated values of θ
using 7 different estimators

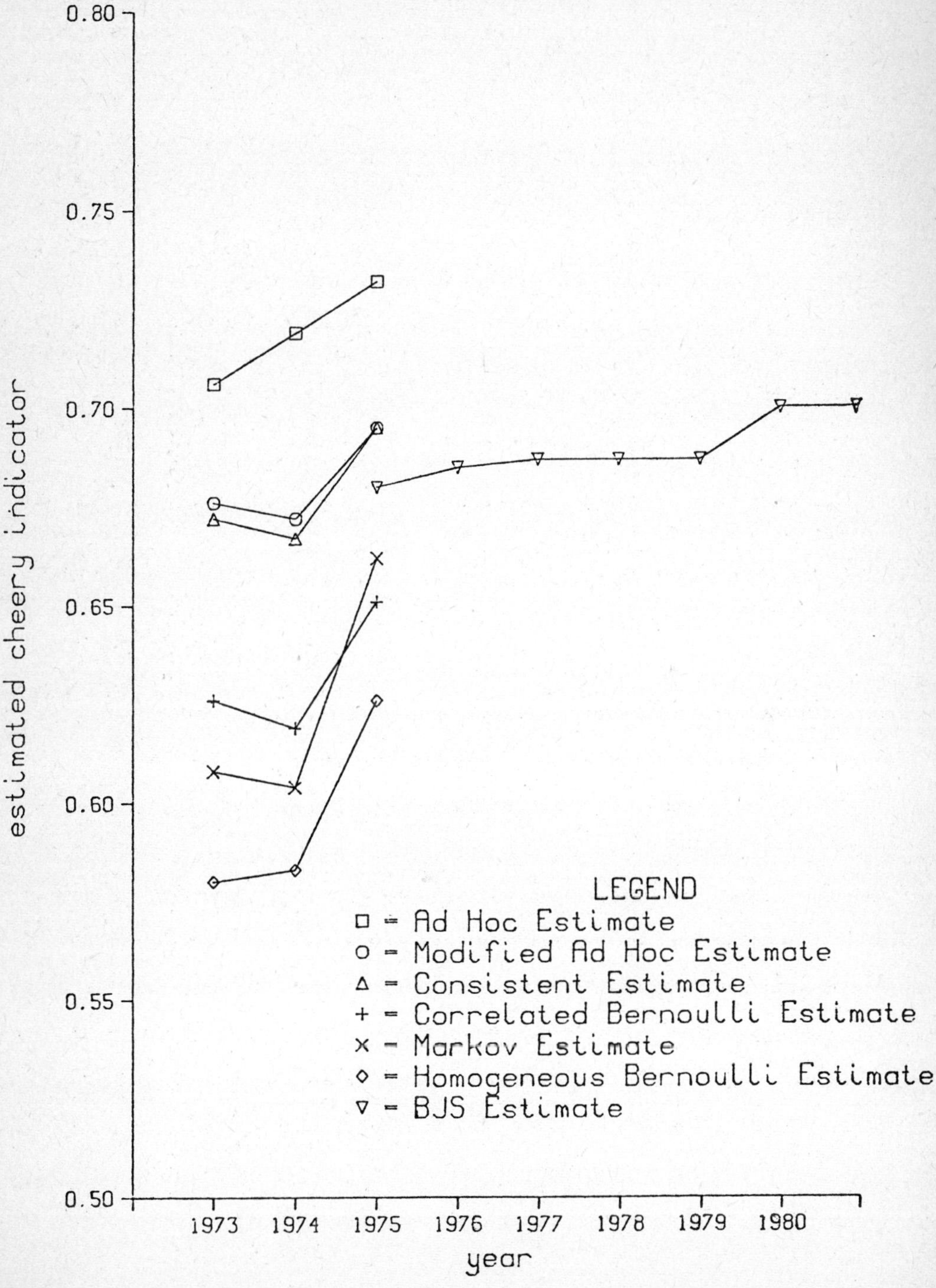

From Figure 12 we see that the homogeneous Bernoulli model, whose fit was not very good, produces estimates that are well below those of the better fitting models. This would seem to indicate that victimization does not occur independently from month to month but rather is positively correlated. We see another indication of positive correlation from the Markov model. For 1975 the estimated transition matrix is

$$\begin{pmatrix} .967 & .033 \\ .838 & .162 \end{pmatrix}$$

If there were independence from month to month, we would expect the first row of this matrix to be very close to the second row. Using the information in Table 14, we find that the difference between $\hat{p}_{00}$ and $\hat{p}_{10}$ is .129 with an estimated standard deviation of .037. Thus it appears that HUs that are victimized at month j are more likely to be victimized at month j+1 than HUs that are crime-free at month j. Kadane (1983) arrives at a similar conclusion after performing a Bayesian analysis.

Figure 13 shows the estimated probability of victimization plotted against the number of months of information contributed for the consistency model, the homogeneous Bernoulli model, and the Markov model using the 1973 data. The observed proportion of victimized HUs is also included on the plot. This same information using the 1975 data is displayed in Figure 14. The observed proportion of victimized HUs is not smooth for either year but is much less variable for the 1975 data. (The 1974 data is also "bumpier" than the 1975 data.) It may be that the starting up of the rotation scheme and inexperience of interviewers caused some additional variation in the 1973 data and that in more recent years there is less of this variability. The fact that the 1975 data is smoother explains why the models fit that data better than the 1973 data. The curve of estimated probabilities of victimization arising from the consistency model, which had the best fits, bends more than the corresponding curves for the other two models. All of the models overestimate the proportion of HUs contributing 12 months of information that are victimized. This may indicate that the HUs that have had at least one nonresponse, and thus have not contributed 12 months of information, are victimized more often than HUs that have had no nonresponse. In the next chapter, we examine this issue in detail.

FIGURE 13

The observed proportion of victimized HUs
and the expected proportion of victimized HUs
under 3 models.
1973 data

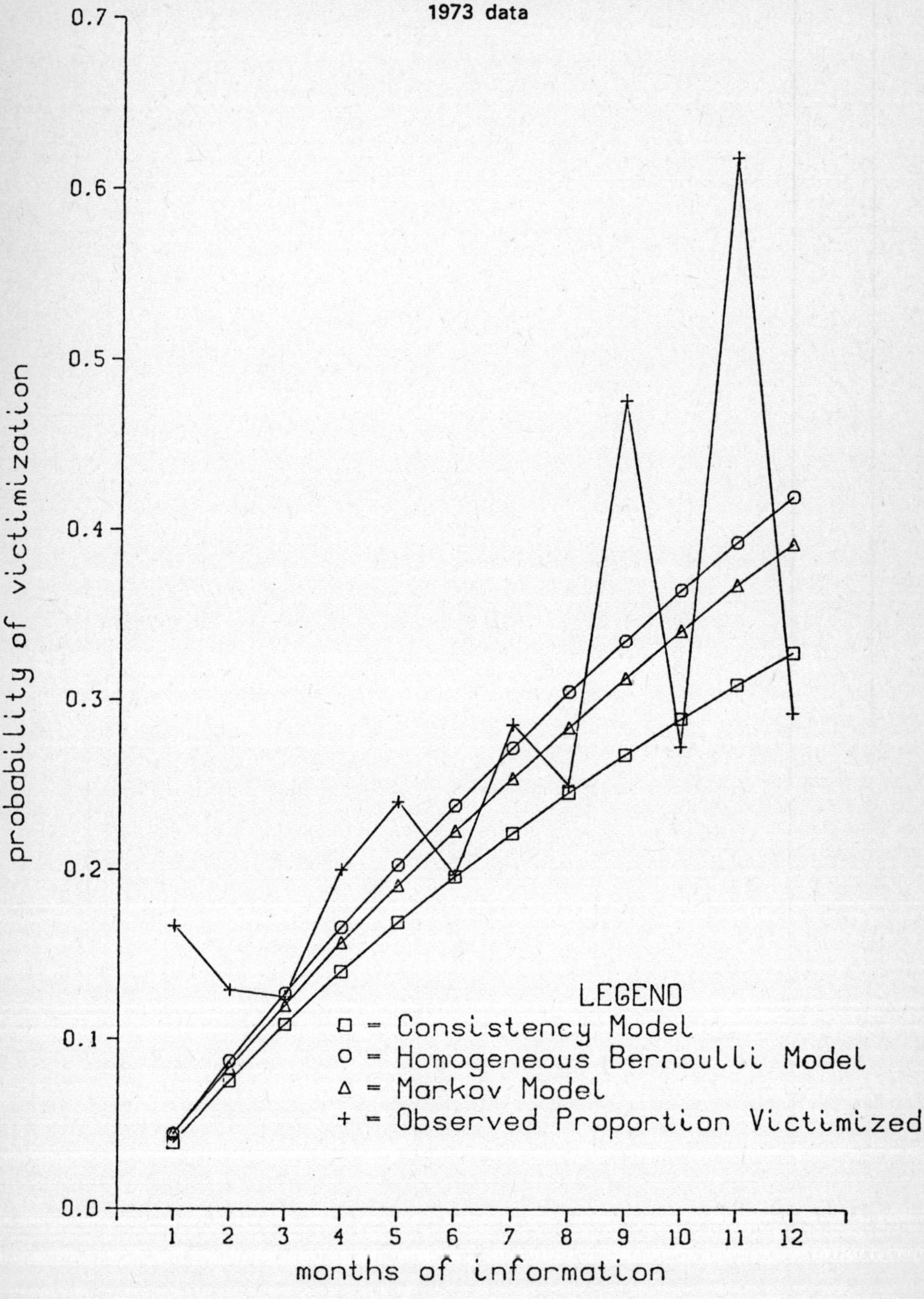

FIGURE 14

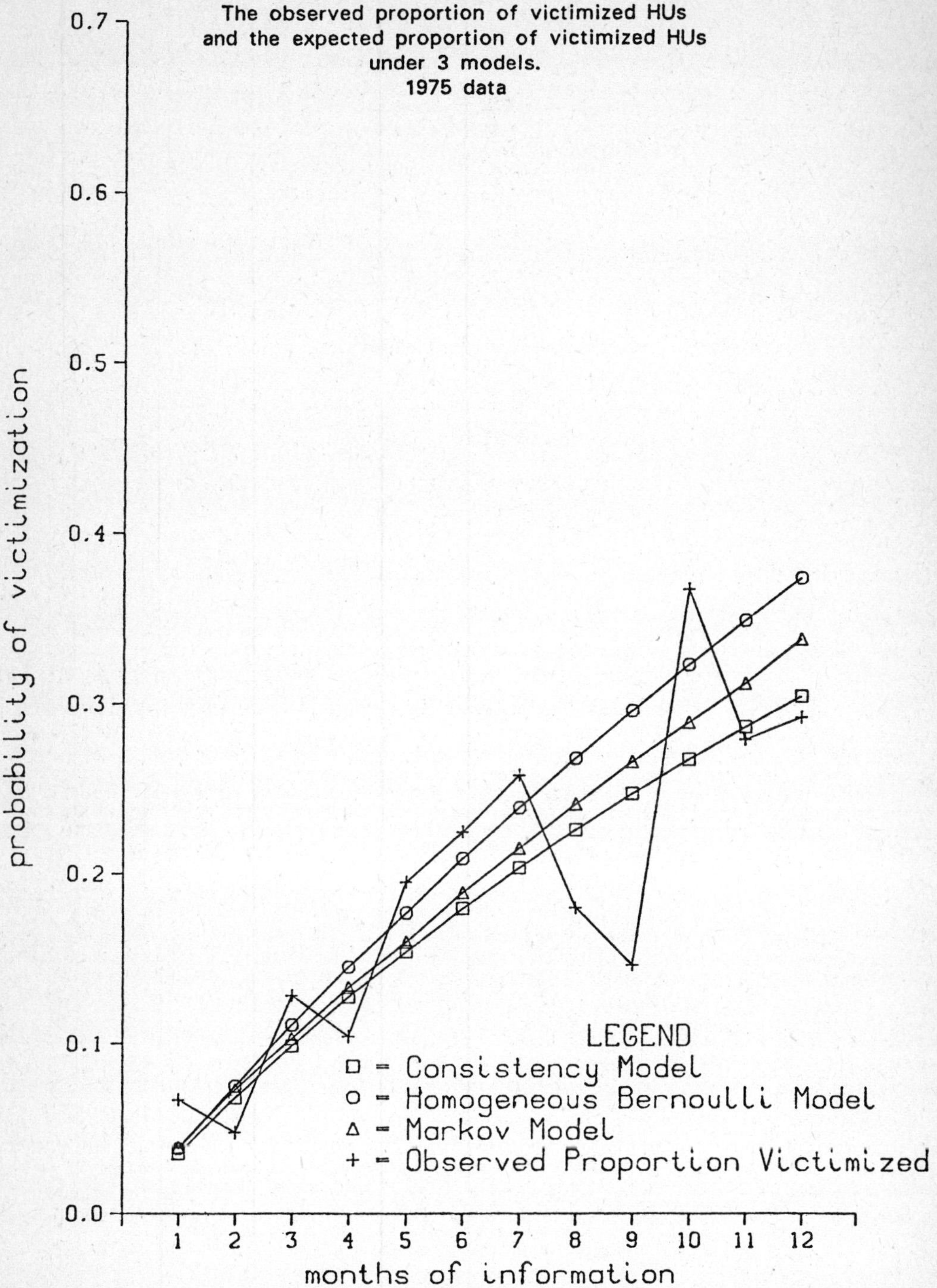
The observed proportion of victimized HUs
and the expected proportion of victimized HUs
under 3 models.
1975 data
0.7
0.6
0.5
0.4
0.3
0.2
0.1
0.0
probability of victimization
1
2
3
4
5
6
7
8
9
10
11
12
months of information
LEGEND
□ = Consistency Model
○ = Homogeneous Bernoulli Model
Δ = Markov Model
+ = Observed Proportion Victimized

CHAPTER 5

NONRESPONSE

In the previous three chapters we discussed several estimators of the prevalence of crime under the assumption that the missing data were missing at random. Here we examine the validity of this assumption and find that it is not supported by the data. HUs that have had at least one month of nonresponse in the year of interest have consistently lower cheery indicators and thus nonresponse and victimization are positively correlated. Because of this positive correlation, modelling of the missing data leads to lower estimates of θ.

5.1. Checking the Missing at Random Assumption

In order to check the missing at random assumption, the HUs in the 1% sample were divided into two groups for each of the years 1973 through 1975. The first group contains HUs that had at least one month of information in the year of interest but no months covered by a noninterview in that year. The second group consists of the HUs that had at least one month of information in the year of interest and for which at least one month of the information for the year was missing due to a noninterview. The first group contains 718, 849, and 822 HUs for 1973 through 1975, respectively, and the second group contains 118, 117, and 133 HUs. The homogeneous Bernoulli model was then fit to each of these two groups and the results are displayed in Table 18. The χ^2 statistics for the HUs with no noninterviews have 10 degrees of freedom while the statistics for the HUs with at least one nonresponse have 9.

This is due to the fact that, since HUs with at least one noninterview cannot contribute 12 months of information, there is one less cell in our table of observed and expected counts.

TABLE 18

Homogeneous Bernoulli Model
for HUs with no noninterviews
and HUs with at least one noninterview

no noninterviews

	$1-\hat{p}$	$\hat{\theta}_3$	est. st. dev.	χ^2
1973	.9587	.6028	.0175	20.66
1974	.9576	.5949	.0171	29.90
1975	.9651	.6528	.0171	14.26

at least one noninterview

	$1-\hat{p}$	$\hat{\theta}_3$	est. st. dev.	χ^2
1973	.9283	.4096	.0476	18.20
1974	.9377	.4624	.0510	13.25
1975	.9364	.4546	.0455	4.65

	difference	est. st.dev.
1973	.1932	.0507
1974	.1325	.0538
1975	.1982	.0486

We see from the χ^2 statistics that the fit of this model for the HUs with no noninterviews is still poor for the years 1973 and 1974. As when we fit this model to all of the data, the fit for 1975 is better than the fits for the previous two years. In Chapter 4 we had speculated that the poor fit of the homogeneous Bernoulli model might be due to the HUs with at least one nonresponse. Here we see that even when the nonresponding HUs are removed, the fit is still not good. The major contributions to these χ^2 statistics still come from the HUs with 12 months of information for the year. For these HUs, the probability of being crime-free is always underestimated.

For the HUs with at least one noninterview month, the homogeneous model seems to fit better, and, as we saw before, the 1975 fit is the best. The values of $\hat{\theta}_3$ calculated from these HUs are all lower than the corresponding estimates computed from the HUs with no nonresponse.

The last section of Table 18 shows the difference ($\hat{\theta}_3$ for the HUs with no nonresponse minus $\hat{\theta}_3$ for the HUs with at least one nonresponse) and the estimated standard deviation for each of the three years. Since the model does not appear to fit well for the 1973 and 1974 data, the estimates for those years may not be very meaningful, but for the 1975 data the difference is significant with the nonresponse HUs having a lower cheery indicator than the HUs with no nonresponse.

For these same two groups of HUs the Markov model has been fit for each of the three years. The results are shown in Table 19. Here, the χ^2 statistics for the HUs with no noninterviews have 8 degrees of freedom while those for the HUs with at least one noninterview have 7. For 1973 the Markov model fits the data from the HUs with no nonresponse fairly well but the fit for the HUs with a nonresponse is poor. For 1974 we have the opposite situation -- the model fits the data for the HUs with nonresponse better than the data for the HUs with no nonresponse. The model seems to fit both groups of the 1975 data fairly well as with the homogeneous Bernoulli model.

TABLE 19

Markov Model
for HUs with no noninterviews
and HUs with at least one noninterview

no noninterviews

	$\hat{\theta}_4$	est. st. dev.	χ^2
1973	.6294	.00027	11.31
1974	.6194	.00079	24.71
1975	.6850	.00081	9.59

at least one noninterview

	$\hat{\theta}_4$	est. st. dev.	χ^2
1973	.4317	.0030	39.16
1974	.4602	.0034	16.16
1975	.4907	.0029	13.24

	difference	est. st.dev.
1973	.1977	.0030
1974	.1592	.0035
1975	.1943	.0030

When all of the data were used in the estimation, the 1973 χ^2 value was 29.70 and so the fit has been substantially improved by considering only the HUs with no nonresponse. For 1974 and 1975 the changes in χ^2 are much smaller. Thus we have some evidence to believe that the poor fit of the Markov model to the 1973 data was at least partly due to the HUs with nonresponse. This does not seem to be the case for the 1974 data. Again we see that the HUs with no nonresponse yield higher estimates of θ than the HUs with nonresponse.

The same two groups of HUs have been used to estimate θ from the consistency model. The results of these computations are displayed in Table 20. The χ^2 statistics for the HUs with no noninterviews have 10 degrees of freedom while the statistics for the HUs with at least one noninterview have 9. As when we fit this model to all of the data, the fit is good for all three years. Here we see that the model seems to perform well for the HUs with nonresponse as well as for the HUs without nonresponse. Again the HUs with nonresponse produce estimates of θ that are lower than the estimates using only HUs with no nonresponse. For this model we find that the estimates of θ are higher for each group of data and for each year than the corresponding estimates for the homogeneous Bernoulli model and the Markov model. The differences between the estimates using only the HUs with no nonresponse and the estimates using the nonresponse HUs are smaller than the differences found for the other two models. Still, we have fairly strong evidence that these differences are positive. For 1973, the difference is 2.98 times the estimated standard deviation, for 1974, 3.94 times, and for 1975, 6.13 times.

Thus we find that, even when we remove HUs with nonresponse, the Bernoulli model does not seem to fit the data. For the Markov model, removal of the HUs with nonresponse improves the fit but only for the 1973 data. The consistency model still describes the data best, for HUs with nonresponse and for HUs without nonresponse.

The data do not support the assumption that the missing data are missing at

TABLE 20

Consistency Model
for HUs with no noninterviews
and HUs with at least one noninterview

no noninterviews

	$\hat{\theta}_2$	est. st. dev.	χ^2
1973	.6923	.0183	2.36
1974	.6820	.0099	5.13
1975	.7161	.0067	1.57

at least one noninterview

	$\hat{\theta}_2$	est. st. dev.	χ^2
1973	.5385	.0486	5.27
1974	.5553	.0306	5.97
1975	.5687	.0231	1.35

	difference	est. st.dev.
1973	.1548	.0519
1974	.1267	.0322
1975	.1474	.0241

random. HUs with at least one month covered by a nonresponse in the year of interest tend to have significantly lower cheery indicators than the HUs with no nonresponse. Thus nonresponse and victimization tend to be positively correlated.

Recall that a type A nonresponse occurs, basically, when the residents of an HU are not contacted, a type B noninterview occurs when the HU is vacant, and a type C noninterview occurs when the HU does not exist. Because these three situations are rather different, we may wonder if victimization differs for HUs with different types of nonresponse. To answer this question, we would like to look at the victimization records of HUs that have had each particular type of nonresponse. Several problems arise when the 1% subsample is divided into groups according to type of nonresponse. First, some HUs have more than one type of nonresponse in the year of interest. Second, as we saw in Table 3, for any particular year there are relatively few HUs with A, C, or unknown type nonresponses and thus it is difficult to determine whether those HUs were victimized differently from the others. To get at least some indication of whether or not victimization varies with type of nonresponse, the HUs in the 1% subsample have been grouped for each of the three years by type of nonresponse, with an HU being classified in the, for instance, type A group if it had at least one month of type A noninterview in the year of interest. Thus an HU may appear in more than one of the four groups. For each of the groups, we find the number of HUs with no information in the year of interest. These HUs had only nonresponses for the year. These counts appear in the 4th column of Table 21. From the remaining HUs we total the number of crime-free months and the number of victimization months. These numbers appear in the 1st and 2nd columns of Table 21, and the 3rd column contains the proportion of crime-free months.

From this table we see that there are too few months in HUs with unknown types of nonresponse and type C nonresponse to make any statements about the victimization in such HUs. Between the other two types of nonresponse, we find the same pattern in each of the three years: the type B nonresponse group has the highest proportion of victimization months, followed by the type

TABLE 21

Victimization months by type of nonresponse

	crime-free months	vict. months	proportion crime-free	HUs with no info
1973				
no nonrsp.	7122	308	.041	--
A	242	16	.062	15
B	428	42	.089	104
C	37	3	.075	22
U	14	1	.067	1
1974				
no nonrsp.	7398	326	.042	--
A	266	9	.033	20
B	412	38	.084	113
C	12	1	.077	30
U	14	0	.000	2
1975				
no nonrsp.	7342	269	.035	--
A	261	10	.037	23
B	528	42	.074	114
C	30	1	.032	17
U	55	4	.068	4

A group. For each of the three years, the HUs with no nonresponse and the HUs with type A nonresponse have lower proportions of victimization months than the HUs in the type B group. In fact, the proportion of victimization months for HUs with type B nonresponse is at least twice that for HUs with no nonresponse. This is a rather interesting observation in light of the models derived in Chapter 3 under which the BJS estimators are consistent. Recall that, in these models, an HU that has one type B or C nonresponse and one interview has probability 1 - θ of reporting a victimization at that interview. This is the same probability that an HU that is interviewed twice reports at least one victimization. Under such a model, HUs with type B or C noninterviews have the same probability of being victimized in 6 months as HUs with two interviews have of being victimized in 12 months. In addition, under these models for which the BJS estimators are consistent, the probability that an HU with an interview and a type A nonresponse is victimized is the same as the probability that an HU with two interviews reports a victimization at one particular interview. Thus HUs with type B and C nonresponses have a higher probability of being victimized during any particular month than HUs with type A or no nonresponses. Judging from the information about type B nonresponses in Table 21, such a model may provide a fairly accurate description of victimization. *Thus the BJS estimators may, accidentally, take into account differences in victimization associated with type A and B nonresponse.*

Although, because of the problems of multiple types of nonresponse in one HU and rarity of type A, C, and U noninterviews, it is difficult to make meaningful comparisons among the numbers in Table 21, HUs with different types of nonresponse tend to have different proportions of victimization months. Notably, HUs with type B noninterviews have rather high proportions of victimization months relative to HUs with type A nonresponses. In addition, the majority of HUs with only nonresponse in the year of interest have type B nonresponses. Since, at this point, we do not have sufficient evidence to conclude that victimization rates are significantly different for the different groups, we continue to combine all nonresponses. However, we note that analyses with larger data sets should be performed to determine whether or not the differences are indeed significant.

5.2. Taking the Nonresponse into Account

One way to account for nonresponse in a model is to include the number of interviews during the year of interest for which the HU did not respond as an explanatory variable. We have done this by using a logit model to describe the probability of being victimized at least once during the part of the year for which an HU is interviewed as a function of:

a. the number of months of information they contribute to the year (1-12)

b. the number of interviews for which they were nonrespondents for that year (0-2)

c. the number of interviews for which they were in the sample (1-3)

A table of these three explanatory variables will necessarily have many structural zeros because of the restrictions that certain values of any one of the variables place on the possible values of the other two. In Table 22 below, X's appear in the cells that contain structural zeros. There are 31 cells that do not contain structural zeros. For 1975, the cross-classification of the 1% sample is shown in Table 23.

We cannot fit a logit model to these data as they are since one of the observed proportions of victimized HUs is 0/0. For a few simple models, we have tried 2 different methods of handling this zero:

a. put weight zero on the cell (thereby losing a degree of freedom)

b. treat the cell count as 0/.5

TABLE 22

X's represent structural zeros due to the sampling and data collection schemes

In sample for 1 interview :

		months of information											
		1	2	3	4	5	6	7	8	9	10	11	12
	0							X	X	X	X	X	X
nonints	1	X	X	X	X	X	X	X	X	X	X	X	X
	2	X	X	X	X	X	X	X	X	X	X	X	X

In sample for 2 interviews :

		months of information											
		1	2	3	4	5	6	7	8	9	10	11	12
	0	X	X	X	X	X	X						
nonints	1							X	X	X	X	X	X
	2	X	X	X	X	X	X	X	X	X	X	X	X

In sample for 3 interviews :

		months of information											
		1	2	3	4	5	6	7	8	9	10	11	12
	0	X	X	X	X	X	X	X	X	X	X	X	
nonints	1	X	X	X	X	X							X
	2							X	X	X	X	X	X

TABLE 23

1975 information
HUs victimized/# HUs in cell

In sample for 1 interview :

# of nonints	1	2	3	4	5	6	7	8	9	10	11	12
	months of information											
0	2/27	2/41	5/32	4/41	6/34	6/39						
1												
2												

In sample for 2 interviews :

# of nonints	1	2	3	4	5	6	7	8	9	10	11	12
	months of information											
0							4/21	7/39	3/34	10/33	6/29	24/82
1	0/1	0/0	0/3	0/3	1/3	6/28						
2												

In sample for 3 interviews :

# of nonints	1	2	3	4	5	6	7	8	9	10	11	12
	months of information											
0												108/370
1						7/17	4/10	2/11	3/7	8/16	6/14	
2	0/2	0/1	0/4	1/4	1/4	1/5						

The results of these two methods were very similar and we opted to use the second.

The G^2 statistics and degrees of freedom obtained using GLIM[3] for several models are given below. MONTH represents the number of months of information that the HU contributed for the year and ranges from 1 to 12; INTSHD represents the number of interviews that the HU should have had, i.e., the number of interviews for which they were in the sample, and ranges from 1 to 3; NONINT represents the number of noninterviews the HU had during the year and ranges from 0 to 2.

model	G^2	df
1) mean only	60.38	30
2) MONTH	18.48	19
3) INTSHD	29.10	28
4) NONINT	54.39	28
5) MONTH, INTSHD	13.77	17
6) MONTH, NONINT	10.52	17
7) INTSHD, NONINT	25.05	26
8) MONTH, INTSHD, NONINT	9.60	15
9) MONTH*NONINT	1.97	2

Models 1 and 4 clearly do not fit well, while the fits of the other models appear to be good. Comparing models 2 and 5, we find a difference in G^2 of 4.71 on 2 degrees of freedom. The INTSHD term is thus marginally significant. Comparing models 2 and 6, we find a difference in G^2 of 7.96 on 2 degrees of freedom and conclude that there is an effect due to NONINT.

[3] Generalized Linear Interactive Modelling (GLIM) is a project sponsored by the Working Party on Statistical Computing of the Royal Statistical Society and developed by R.J. Baker, M.R.B. Clarke and J.A. Nelder, Rothamsted Experimental Station, Harpenden, Herts, England.

Thus we will retain models 5 and 6 but not model 2. The difference in G^2 statistics for models 3 and 5 is 15.33 and, since this difference has 11 degrees of freedom, we do not have strong evidence that the month effect is non-zero. Because of this information, we drop model 5 from consideration. Comparing models 3 and 7 we do not find strong justification for adding the NONINT term to model 3 and thus drop model 7. The comparisons of models 6 and 8 and models 6 and 9 indicate that there is no need to add a third term. Thus we are left with a choice between models 3 and 6:

3) INTSHD	G^2=29.10	df=28
6) MONTH, NONINT	G^2=10.52	df=17

Note that the second of these two models is much more intuitive than the first. It's hard to understand why the probability of being victimized should depend on the number of interviews an HU *should* have had in the year. Because of this interpretation problem, we will not consider model 3 any further. The standardized residuals for model 6 are shown in Table 24 and we see that they do not exhibit any trends.

Since the explanatory variables have ordered categories, we may be able to reduce the number of parameters needed to obtain a good fit by using ordered category models. In the following tables, an "S" before a variable's name indicates that scores were used for that category. The scores for MONTH are -11, -9, -7, ..., 11; for NONINT they are -1, 0, 1.

model	G^2	df
10) SMONTH, NONINT	17.65	27
11) SMONTH, SNONINT	21.22	28

By using scores for the MONTH variable, we are able to save 10 degrees of freedom while preserving a good G^2 value. Model 11, in which scores are

TABLE 24

Standarized Residuals for the model $\log\frac{p_{ijk}}{1-p_{ijk}} = \mu + \mu_{NONINT(j)} + \mu_{MONTH(k)}$

$i=1,2,3;\ j=0,1,2\ ;\ k=1,2,\ldots,12$

In sample for 1 interview :

# of nonints	months of information 1	2	3	4	5	6	7	8	9	10	11	12
0	.22	.08	.66	.02	-.08	-.05						
1												
2												

In sample for 2 interviews :

# of nonints	months of information 1	2	3	4	5	6	7	8	9	10	11	12
0							-.20	.48	-.68	-.06	-.25	.01
1	-.38	-.23	-.92	-.84	.03	-.86						
2												

In sample for 3 interviews :

# of nonints	months of information 1	2	3	4	5	6	7	8	9	10	11	12
0												-.01
1						1.13	.24	-.72	1.17	.07	.31	
2	-.41	-.24	-.80	.84	.18	.08						

used for both the month and noninterview categories, also appears to fit well. Since, as we shall see, we can extrapolate using the parameter estimates from model 11 to estimate the probability that an HU with three noninterviews was victimized, we restrict our attention to this model. The standardized residuals for this model appear in Table 25. The fit seems to be fairly good (although the model may have a slight tendency to overestimate the probability of victimization for HUs with two noninterviews).

TABLE 25

Standardized Residuals for the model $\log\frac{p_{ijk}}{1-p_{ijk}} = \mu+(j-1)\mu_{NONINT(j)}+(2k-3)\mu_{MONTH(k)}$

$i=1,2,3;\ j=0,1,2\ ;\ k=1,2,\ldots,12$

In sample for 1 interview :

# of nonints	months of information 1	2	3	4	5	6	7	8	9	10	11	12
0	.04	-.80	1.09	-.25	.89	.18						
1												
2												

In sample for 2 interviews :

# of nonints	months of information 1	2	3	4	5	6	7	8	9	10	11	12
0							.34	-.09	-1.74	.91	-.71	-.06
1	-.37	-.28	-.75	-.81	.55	-.21						
2												

In sample for 3 interviews :

# of nonints	months of information 1	2	3	4	5	6	7	8	9	10	11	12
0												-.16
1						1.77	1.02	-.79	.60	1.20	.28	
2	-.71	-.54	-1.17	.15	-.28	-.70						

Thus the model:

$$\log \frac{p_{jk}}{1-p_{jk}} = \mu + v_j \mu_{NONINT} + w_k \mu_{MONTH} \tag{5-1}$$

seems to adequately describe the data. Here the v_j's are the NONINT scores, the w_k's are the MONTH scores, and the "i" subscripts have been dropped since the variable INTSHD does not appear in the model. Since the NONINT scores are -1, 0, 1 and the MONTH scores are -11, -9, ..., 11, we can rewrite this model as

$$\log \frac{p_{jk}}{1-p_{jk}} = \mu + (j-1)\mu_{NONINT} + (2k-13)\mu_{MONTH} \, . \tag{5-2}$$

Here i ranges form 0 to 2 and k ranges form 1 to 12. The parameter estimates and standard errors for this model are given in Table 26.

TABLE 26

Estimates for model $\log \frac{p_{jk}}{1-p_{jk}} = \mu + (j-1)\mu_{NONINT} + (2k-13)\mu_{MONTH}$

parameter	estimate	est. st. dev.
μ	-1.127	.167
μ_{NONINT}	.583	.190
μ_{MONTH}	.077	.014

Using this model, we can estimate the probability that an HU, with j noninterviews and for which we have k months of information, is victimized at least once during those k months by:

$$\hat{p}_{j,k} = \frac{e^{-1.127+(j-1).583+(2k-13).077}}{1+e^{-1.127+(j-1).583+(2k-13).077}}$$

$$= \frac{e^{-2.706+j(.583)+k(.153)}}{1+e^{-2.706+j(.583)+k(.153)}} \tag{5-3}$$

For no noninterviews and 12 months of information, we find $\hat{p}_{0,12} = .296$. Thus we estimate the probability that an HU with no noninterviews and 12 months of information is crime-free to be .704. Similarly, the probability that an HU with 1 noninterview and 11 months of information is not victimized during those 11 months is estimated to be .608. Notice that this is the estimated probability of the HU being crime-free during the 11 months of information that it contributes, not during the entire year. To estimate the probability that an HU with one noninterview would be victimized in 12 month's time had we actually been able to collect all of the data, we can compute $\hat{p}_{1,12}$:

$$\hat{p}_{1,12} = \frac{e^{-2.706+.583+12(.153)}}{1+e^{-2.706+.583+12(.153)}}$$

$$= .416 \ . \tag{5-4}$$

Thus we can estimate the probability that an HU with one noninterview would not have been victimized in 1975, had we been able to obtain the full year's data for them, to be 1-.416 = .584. Similarly for an HU with two noninterviews, we estimate the probability of being crime-free for the year by $1 - \hat{p}_{2,12} = .453$. The results are summarized in Table 27.

TABLE 27

Results of logit model fit

# nonints.	p(crime-free for 12 months)	est. st. d.
0	.704	.024
1	.584	.046
2	.453	.089

In order to assess the variability of these estimates, we have used the δ method (Bishop et al., 1975) to calculate the estimated standard deviations shown in Table 27. We know that, since the μ's can be expressed as linear combinations of the log expected cell counts, $N^{1/2}(\hat{\mu}, \hat{\mu}_{NONINT}, \hat{\mu}_{MONTH})$ will be asymptotically normally distributed with a certain variance-covariance matrix, say, $\sum$. Thus $N^{1/2}\log[\hat{p}_{jk}/(1-\hat{p}_{jk})]$ will be asymptotically normally distributed with variance

$$\mathrm{Var}\left[N^{1/2}\log\left(\frac{\hat{p}_{kj}}{1-p_{jk}}\right)\right] = (1,\ j-1,\ 2k-13) \sum \begin{pmatrix} 1 \\ j-1 \\ 2k-13 \end{pmatrix}. \tag{5-5}$$

To get the asymptotic variance of $\hat{p}_{jk}$, we can again use the δ method. First we need a function f such that

$$f\left(\log\frac{\hat{p}_{jk}}{1-p_{jk}}\right) = \hat{p}_{jk}. \tag{5-6}$$

This condition is satisfied by $f(x)=e^x/(1+e^x)$. Thus $N^{1/2}\hat{p}_{jk}$ will have asymptotic variance:

$$\mathrm{Var}\left[N^{1/2}\log\left(\frac{\hat{p}_{jk}}{1-p_{jk}}\right)\right]\left[f'\left(\log\frac{\hat{p}_{jk}}{1-p_{jk}}\right)\right]^2 . \tag{5-7}$$

Since $f'(x) = e^x/(1+e^x)^2$, we find that (5-7) becomes

$$(1,\ j-1,\ 2k-13) \sum \begin{pmatrix} 1 \\ j-1 \\ 2k-13 \end{pmatrix} [\hat{p}_{jk}(1-\hat{p}_{jk})]^2 . \tag{5-8}$$

Thus an estimate of the standard deviation of $\hat{p}_{jk}$ is

$$\left[(1,\ j\text{-}1,\ 2k\text{-}13) \left(N^{-1/2}\hat{\sum}\right) \begin{pmatrix} 1 \\ j\text{-}1 \\ 2k\text{-}13 \end{pmatrix} \right]^{1/2} \hat{p}_{jk}(1-\hat{p}_{jk}). \tag{5-9}$$

GLIM provides the following estimate of $N^{-1/2}\sum$:

$$\begin{pmatrix} .02710 & & \\ .02259 & .03602 & \\ -.00039 & .00103 & .00019 \end{pmatrix}$$

and we have used this estimate to produce the estimated standard deviations shown in Table 27.

One way we could combine these results to come up with a single θ (the probability of being crime-free for the year) is to weight each value by the proportion of HUs with the appropriate number of nonresponses. Using this method with the data in Table 21 yields:

$$\hat{\theta}_{logit1} = \frac{822}{955}(.704) + \frac{113}{955}(.584) + \frac{20}{955}(.453) = .685\ . \tag{5-10}$$

However, there are HUs that should have contributed information in 1975 but did not because they had only noninterviews. These HUs do not appear in Table 21. If we include the 102 HUs that had one or two noninterviews and no completed interviews in our weights we find:

$$\hat{\theta}_{logit2} = \frac{822}{1057}(.704) + \frac{169}{1057}(.584) + \frac{66}{1057}(.453) = .669\ . \tag{5-11}$$

In addition, there are 48 HUs that should have had 3 interviews but had 3 noninterviews instead. We can estimate the probability that such an HU would be victimized in the year by $\hat{p}_{3,12}$:

$$\hat{p}_{3,12} = \frac{e^{-2.706+3(.583)+12(.153)}}{1+e^{-2.706+3(.583)+12(.153)}} = .707\ . \tag{5-12}$$

Adding these HUs into our weighted average we find

$$\hat{\theta}_{\text{logit3}} = \frac{822}{1105}(.704) + \frac{169}{1105}(.584) + \frac{66}{1105}(.453) + \frac{48}{1105}(1-.707)$$
$$= .653 . \qquad (5\text{-}13)$$

There are several points that should be raised about the inclusion of these HUs that have nothing but nonresponses for the year. First, we have supposed, in using equation (5-11), that HUs with one or two noninterviews and no completed interview are victimized like HUs with one or two noninterviews and at least one completed interview. We have no means of verifying such an assumption since the HUs in question contribute no data, at least not in the year of interest. Secondly, there is no way to verify that the extrapolation in equation (5-12) is appropriate. Since no HU with three noninterviews in the year of interest can possibly contribute any information in that year, we have no data with which to check this extrapolation. Still, these estimates do provide us with some information. Since $\hat{\theta}_{\text{logit1}}$ does not include any HUs that contribute only nonresponse to the year of interest, and since we know that nonresponse is positively correlated with victimization, we expect $\hat{\theta}_{\text{logit1}}$ to be an overestimate of θ. Although the assumption and extrapolation used in computing $\hat{\theta}_{\text{logit2}}$ and $\hat{\theta}_{\text{logit3}}$ cannot be checked with the data set that we have, they are not intuitively unreasonable and they give us some indication of the variability of $\hat{\theta}_{\text{logit}}$ due to the HUs that contribute only nonresponse.

In order to estimate standard deviations for these prevalence estimators, we can follow a procedure very similar to the one we used to find standard deviations for the $\hat{p}_{j.12}$'s. If we let

$$\mathbf{D} = \begin{pmatrix} \hat{p}_{0,12}(1-\hat{p}_{0,12}) & 0 & 0 \\ 0 & \hat{p}_{1,12}(1-\hat{p}_{1,12}) & 0 \\ 0 & 0 & \hat{p}_{2,12}(1-\hat{p}_{2,12}) \end{pmatrix} \quad (5\text{-}14)$$

and

$$\mathbf{T} = \begin{pmatrix} 1 & -1 & 11 \\ 1 & 0 & 11 \\ 1 & 1 & 11 \end{pmatrix} \quad (5\text{-}15)$$

then an estimate of the variance-covariance matrix of $(\hat{p}_{0,12}, \hat{p}_{1,12}, \hat{p}_{2,12})$ is

$$\hat{\sum}_{\hat{p}} = \mathbf{D}\ \mathbf{T}\ N^{-1/2}\hat{\sum}\ \mathbf{T}^t\ \mathbf{D}^t\ .$$

Thus an estimate of the variance of $\hat{\theta}_{logit1}$ is

$$\left(\frac{822}{955}, \frac{113}{955}, \frac{20}{955} \right) \hat{\sum}_{\hat{p}} \begin{pmatrix} 822/955 \\ 113/955 \\ 20/955 \end{pmatrix} . \quad (5\text{-}16)$$

We note that, since there is actually variability in these transformation vectors, these estimates will underestimate the variance of $\hat{\theta}_{logit1}$. The results of this computation and similar computations for $\hat{\theta}_{logit2}$ and $\hat{\theta}_{logit3}$ are shown in Table 28. Using an analogous method, we can estimate the standard deviations of the differences of the three estimators and these results are also included in Table 28.

We note that even though the range of the estimates is fairly small, the differences in Table 28 are rather large relative to the estimated standard deviations. This, of course, is due to the fact that the $\hat{\theta}_{logit}$'s are highly correlated.

We would like to perform similar analyses on the 1974 and 1973 data. Unfortunately, the models that fit the 1975 data do not perform as well with

TABLE 28

Prevalence estimates using a logit model
1975 data

	estimate	est. st.dev.
$\hat{\theta}_{logit1}$	.685	.021
$\hat{\theta}_{logit2}$	.669	.023
$\hat{\theta}_{logit3}$	.653	.026

	difference	estimated st. dev.
$\hat{\theta}_{logit1}-\hat{\theta}_{logit2}$	.015	.0058
$\hat{\theta}_{logit2}-\hat{\theta}_{logit3}$	.016	.0046
$\hat{\theta}_{logit1}-\hat{\theta}_{logit3}$	.032	.0103

the 1974 data. The model that includes the categories MONTH and NONINT has a G^2 value of 24.32 with 17 degrees of freedom, but when we fit the ordered category model that includes the terms SMONTH and SNONINT we find that G^2 is 52.90 with 28 degrees of freedom -- a very poor fit. In fact, none of the ordered category models describe the 1974 data well.

For the 1973 data the ordered category model that includes the SMONTH and SNONINT terms has a G^2 value of 39.81 with 28 degrees of freedom. This is a marginal fit but the residuals do not seem to exhibit any trends. The tendency that we noticed in the 1975 residuals for this model to overestimate the probability of victimization in HUs with two nonresponses does not appear in the 1973 residuals. In addition, there are two cells whose combined contribution to a χ^2 goodness-of-fit statistic is more than 12. Among the 11 HUs with two interviews, no noninterviews, and 11 months of information, 7 are victimized. The standardized residual for this cell is 2.649. There is only one HU with one interview, one noninterview, and one month of information and this HU has been victimized. The standardized residual for this cell is 2.245. All of the other standardized residuals are much smaller than these two. Thus, it appears that there are two rather unusual proportions of victimized HUs in the 1973 data but that, apart from these two cells, the ordered category model that we used to describe the 1975 data fits the 1973 data well.

The results for 1973 of calculations similar to those done for the 1975 data are displayed in Table 29. As with the 1975 estimates we find that the range of values is fairly small and as we have found with all of the other models, the 1973 estimates are lower than the 1975 estimates.

TABLE 29

Prevalence estimates using a logit model
1973 data

	estimate	est. st.dev.
$\hat{\theta}_{logit1}$	.673	.020
$\hat{\theta}_{logit2}$	.659	.023
$\hat{\theta}_{logit3}$	.641	.026

For 1973, the only models that we have fit that seem to accurately describe the data are the model for which the modified ad hoc estimator is consistent and the logit model. The prevalence estimate based on the consistency model was .672 which is very close to $\hat{\theta}_{logit1}$ and slightly higher than $\hat{\theta}_{logit2}$ and $\hat{\theta}_{logit3}$. Thus the model that includes nonresponse produces one estimate that is about the same and two that are slightly lower than the only other one of our models that has an acceptable fit. The Bernoulli, Markov, consistency and logit models have all produced acceptable fits to the 1975 data. The corresponding estimates range from .626 for the Bernoulli model and .695 for the consistency model. We note that the Bernoulli model is actually a special case of the Markov model. In particular if $p_{00} = p_{10} = p_0$ then the Markov model reduces to the Bernoulli model. In Chapter 4, we found for the 1975 data that p_{00} was significantly different from p_{10} indicating that the Markov model is an improvement over the Bernoulli model. With this in mind, we may want to drop the .626 estimate from consideration. In this case, our acceptable estimates of θ for 1975 range from $\hat{\theta}_{logit3} = .653$ to the consistency estimate $\hat{\theta}_2 = .695$. The Markov estimate, .662, is very close to $\hat{\theta}_{logit2}$.

In summary, we have found that the HUs with at least one nonresponse in the year of interest have significantly lower estimated cheery indicators than HUs with no nonresponse. An ordered category logit model that includes the two explanatory variables "months of information" and "number of noninterviews" provides a good fit to the 1973 and 1975 data. The resulting overall estimates do not differ substantially from the estimates computed from the models discussed in Chapters 3 and 4.

We note at this point that these analyses should not be considered as complete analyses of the data set. They ignore several biases and correlations that are known to exist in the NCS. It is hoped that they will provide a basis from which more accurate models may be developed. In the next chapter we discuss some possible extensions to the models that have been presented.

CHAPTER 6

FURTHER RESEARCH AND CONCLUSIONS

6.1. Extensions

In this section, we present a preliminary analysis of estimation at the HH level and discuss several possibilities for further research.

6.1.1. Estimation at the HH Level

All of the estimation in the previous chapters has been performed at the HU level and thus we have been estimating the probability that an HU is crime-free for the year. A more interesting and intuitive quantity to estimate is the probability that an HH is crime-free for the year. As discussed in Chapter 1, there are several problems associated with estimation at the HH level. The NCS is a survey of housing units and not households. When a family in the survey moves, no attempt is made to relocate this household. Instead, the household that moves into the unit is interviewed. Because of this, many interviews are left unbounded and incidents that occurred earlier than the six months preceding the interview may be telescoped into the reference period. An additional problem arises when a noninterview occurs between two interviews for two different HHs in the same HU. If we want to include information about this noninterview in our model as we did in Chapter 5, we do not know whether or not to attribute the nonresponse to the first or second HH. In this section we do some preliminary analysis to determine the sensitivity of the logit estimates to different ways of assigning the nonresponses to the HHs. We do not try to adjust for telescoping of events during unbounded interviews but we note that it is a problem that should be researched.

For the 1975 data we have fit the ordered logit model with explanatory variables "months of information" and "number of noninterviews" using two different rules to assign nonresponses to HHs. First we ignore all noninterviews that occur between HHs. If a noninterview occurs at the beginning of 1975, we check the 1973 and 1974 information to determine if this noninterview has occurred between HHs or within one HH. If this noninterview occurs in the first record for the HU, it is ignored. If a noninterview occurs at the end of 1975, it is ignored. Thus only nonresponses that we are sure occurred within one HH are considered.

When we fit the ordered category logit model to the data, we find a G^2 value of 19.28 with 28 degrees of freedom but the noninterview term is not significantly different from zero as we see from the estimates and estimated standard deviations displayed in Table 30. The actual estimate itself is negative, indicating lower probabilities of victimization for HHs with more noninterviews. These unexpected results are due to the fact that, by including only the noninterviews that we are certain have occurred within one HH, we have eliminated a large number of the noninterviews. Only 42 HHs have one noninterview (8 are victimized) and only 2 have two noninterviews (neither of these is victimized). Since we used this model at the HU level, we compute an analogous HH estimate using this same model in addition to computing an estimate based on the ordered category model that includes only the "months of information" variable. The results of the fit of this model are displayed in Table 31. Any HH that contributes no information in 1975 is, because of the way in which we assign nonresponses to HHs, treated as though it were out of sample. Thus we do not compute values of $\hat{\theta}_{\text{logit2}}$ and $\hat{\theta}_{\text{logit3}}$ for these two models.

TABLE 30

Ordered categories logit model with explanatory variables "months of information" and "number of noninterviews"

1975 data
Analysis at HH level
Nonresponses between HHs ignored

parameter estimates

	estimate	est. st.dev.
μ	-1.50	.347
μ_{NONINT}	-.057	.357
μ_{MONTH}	.047	.011

$\hat{\theta}_{logit\,1} = .715$

estimated st. dev. = .020

TABLE 31

Ordered categories logit model with explanatory variable "months of information"

1975 data
Analysis at HH level
Nonresponses between HHs ignored

parameter estimates

	estimate	est. st.dev.
μ	-1.44	.090
μ_{MONTH}	.048	.011

$\hat{\theta}_{logit\,1} = .705$

estimated st. dev. = .031

These two estimates of $\theta_{\text{logit 1}}$ are very close together and each is greater than any of the other estimates that we have computed using all of the HUs in the survey. We are still using exactly the same number of months of victimization and number of crime-free months as we were in the estimates of Chapters 2, 3, and 4 (and for this reason, the homogeneous Bernoulli estimates at the HH level would be the same as those at the HU level) but there are several aspects of the data that have changed. First, the average number of months of information that an HH contributes to the year will be smaller than the average for HUs. Secondly, one victimized HU can be 1, 2, or 3 victimized HHs. If, for instance, all of the victimized HUs actually consisted of one victimized HH and one crime-free HH, we would expect the estimate of θ for the HHs to be larger than the estimate for the HUs.

The second rule that we have used to assign nonresponses to HHs is to assign any nonresponses between 2 HHs to both HHs. Nonresponses that occur at the end of 1975 are assigned to the last HH and nonresponses that appear in an HU's first record are attributed to the first HH. Thus, any nonresponse that could possibly have taken place while the members of an HH were living in the unit is assigned to that HH. The ordered category logit model with explanatory variables "months of information" and "number of noninterviews" yields a G^2 value of 24.64 with 28 degrees of freedom. The residuals for this model do not exhibit any trends and the results of the fit are shown in Table 32.

The estimate of $\theta_{\text{logit 1}}$ is very close to the .705 and .715 values computed by ignoring all of the between HH noninterviews. All three of the estimates in Table 32 are greater than the logit estimates computed in Chapter 5 for the 1975 data. Thus it appears, at least in this preliminary analysis, that estimation at the HH level produces slightly higher cheery indicators than

TABLE 32

Ordered categories logit model with explanatory variables "months of information" and "number of noninterviews"

1975 data
Analysis at HH level
Nonresponses between HHs assigned to both HHs

parameter estimates

	estimate	est. st.dev.
μ	-1.17	.167
μ_{NONINT}	.329	.186
μ_{MONTH}	.052	.011
$\hat{\theta}_{logit1}$	.707	.021
$\hat{\theta}_{logit2}$	.696	.023
$\hat{\theta}_{logit3}$	.688	.026

estimation at the HU level. The different ways of assigning the noninterviews to the HHs has not greatly changed the estimates.

6.1.2. Possibilities for Further Research

There are several aspects of the data that have not yet been discussed and which we present here as topics for further research.

First, the National Crime Survey is affected, as are other rotating panel surveys, by rotation group bias. Lehnen and Reiss(1978) have found that the longer an HU is in the survey, the less likely they are to report a victimization. This decrease may be due to any of at least three different reasons. One possible explanation is that respondents may tire of answering the questionnaire and after several interviews may not report an incident that, had it occurred earlier, they would have reported at one of their first interviews. After several exposures to the questionnaire a respondent may not try as hard to recall victimizations that he/she experienced during the previous six months. The respondent may want to finish the interview as quickly as possible and thus may not even report victimizations that he/she remembers.

A second explanation, which is really a special case of interviewee fatigue, is that once a respondent has answered "yes" to one of the screening questions and is then asked a sometimes rather lengthy series of questions about the victimization, he/she may be reluctant to answer "yes" to a screening question at a subsequent interview. According to Lehnen and Reiss' results, this is not the case. In fact, they find that the more crime incidents a respondent has reported, the more likely it is that he/she will report an incident at the current interview.

A third possibility is that rotation group bias is due to HHs that move out of sampled HUs. In order for a respondent to be in the NCS for three or four interviews, he/she must have resided in the same location for several years and at least some member of the HU must have been at home to complete the interview. It may be that this type of person is less likely to be victimized than someone who moves often or is often away from home. In addition, persons residing in the same location for long periods of time may have learned how to best protect themselves and their property in that location. Either or both of these situations would result in the lower victimization rates that we find for people who are in the survey for longer periods of time. To explore the question of whether or not respondents in HUs in which several HHs have lived are more likely to report victimizations, all of the HHs in the 1% subsample that have completed at least one interview have been classified in Table 33 by HH number (first, second, third, at least fourth HH to be interviewed at the HU) and by victimization status at the most current interview (crime-free, household victimization, personal victimization, both household and personal victimization).

TABLE 33

HHs classified by HH number and victimization status

Expected counts appear below observed counts

		c.f.	HH	pers.	both
	1	1135	82	90	44
		1104.6	94.9	103.2	48.2
	2	234	25	31	14
HH#		248.6	21.4	23.2	10.9
	3	65	11	9	4
		72.8	6.3	6.8	3.2
	4	32	8	7	2
		40.1	3.4	3.7	1.7

The expected values under the model of independence appear below the observed counts. We see that as the HH number increases, there are less crime-free and more victimized HHs than anticipated. The total χ^2 value is 25.53 with 9 degrees of freedom and thus we have strong evidence that HH number and victimization status are not independent. HHs in HUs from which families have moved during the HU's time in sample tend to report more victimizations than the first HHs interviewed in sample HUs. Thus it may be that people who move often are more likely to be victims of crime. This would at least in part explain why respondents who are in the sample for longer periods of time are less likely to report victimizations.

The fact that respondents (or HHs) that have been in the survey longer tend to report less crime has not been accounted for in any of our models and is a problem that should be considered in further research. It may be that the slight increase in the cheery indicators from 1973 to 1975 is due to rotation group biases. Since the survey began in 1972, most of the HUs that were interviewed in 1973 were being interviewed for the first, second, or third time and no HU had had more than 4 interviews. By 1975 there were HUs that were being interviewed for the seventh time. For subsequent years, the proportion of HUs in each rotation group should be about the same and we do see a corresponding leveling off in the BJS estimates. Thus the 1973 and 1974 estimates may be slightly lower than subsequent estimates because of the rotation group biases and not because of any change in victimization. We have also noticed, especially in Chapter 4, that the 1973 and 1974 data contain much more variation than the 1975 data. This extra variation may be caused by difficulties in the start-up of the survey. Because of this possibility and the rotation group biases, analyses on data from the years 1976 to the present would be very informative. Unfortunately, longitudinal files for these years are not currently available.

The rotation group bias may also contribute to the fact that so many of our models overestimated the number of HUs contributing 12 months of information that were victimized, even when we restricted our attention to HUs that had no nonresponses in the year of interest. Many of the HUs that do not contribute a full year's information are those HUs that have just been rotated into the sample. During the years 1973 and 1974, no HUs were rotated out of the sample since none had yet finished their 7 interviews and, hence, the HUs with only partial information were being interviewed for the first and second times. Because of the rotation group biases, these HUs may have tended to report more victimizations than expected. By 1975, some HUs were being rotated out of the sample and so some of the HUs with partial information were actually being interviewed for the sixth and seventh times.

Because of the rotation group biases, we would expect these HUs to report relatively few victimizations and perhaps start to "balance out" the HUs that were being interviewed for the first and second times. This theory corresponds with the better fits of the models to the 1975 data. It would be interesting to see if, for later years of data, the Markov model fit was much better than what we have found using the 1973 through 1975 data.

The rotation group bias also provides an possible explanation for the good fit of the logit model in which the probability of victimization depended on the number of interviews that an HU should have had. An HU that should have had only one interview in the year of interest has either just been rotated into the sample or has just been rotated out. We have already discussed the fact that respondents in HUs that have just entered the survey are more likely to report victimizations than those that have been in the survey for a longer period of time. It may be that the probability of reporting a victimization drops off rather quickly as a function of time in survey and then remains fairly constant. If this is the case, the probability of reporting a victimization given that the HU should have had only one interview may not turn out to be the same as the probability that an HU that should have had 2 or 3 interviews reports a victimization.

Another bias that has been observed in the NCS and in other similarly surveys is the recency effect. Respondents do not always remember victimizations that occur to them. In particular, a respondent is more likely to recall a victimization that occurred recently than one that occurred several months ago. In addition, victims sometimes tend to recall incidents as occurring more recently than they actually did occur. These two effects together are known as the recency effect. We have not accounted for this effect in any of the models presented here. We note that modelling this effect using the data currently available would be very difficult at best since we do not know if an incident has been forgotten and we do not know the actual date on which an incident occurred.

Another issue that we have not explored is the addition of covariates to our models. During the interviews, the interviewer collects data on variables that we would expect to be correlated with crime and we should be able to make use of these variables in our modelling. Several problems arise when we consider using these additional variables. First, if we use very many of them, the number of victimized persons, HHs, or HUs in any particular group will become very small making accurate estimation impossible. Secondly, we must be prepared to handle cases in which these background variables are missing. For instance, in our 1% subsample, approximately 7% of the interview records are missing the income variable. A third problem with which we must deal is that the values of these covariates may change from interview to interview thus making longitudinal analyses difficult. Still, we expect the background information to be very valuable in the estimation of θ and methods of incorporating them into longitudinal models should be explored.

One problem that we have avoided here by selecting a subsample consisting of every 100th HU is the clustering problem. All of our models have assumed that the HUs or HHs are victimized independently of one another. Since the HUs are actually chosen in clusters of about 4, in seems unlikely that they actually are victimized independently. Fay (1982a and b) of the Bureau of the Census has developed some procedures for analyzing cross-classified data from complex sample surveys. He uses jackknifing and half-sample methods to measure the underlying variability in estimates obtained by fitting log-linear models. Similar jackknifing procedures could be performed with any of the models presented here. However, since such procedures are generally time-consuming and costly, we would only want to use such a method when we had found a model with which we were fairly well satisfied. It would seem reasonable to develop ways of accounting for rotation group biases and adding covariates to our model before attempting such a large jackknifing project.

All of the results presented here have been computed from a 1% subsample of the data. It would be very informative to perform some of these analyses using the entire data set. It may be that the particular subsample that we have chosen is not representative of the entire sample. As in the case of the jackknifing procedures, such large analyses should be performed for models that we are reasonably sure will accurately describe the data.

6.2. Conclusions

In this monograph we examined several ad hoc estimates to determine under what assumptions they are reasonable estimators of θ, the probability that an HU is crime-free for the year. A very simplistic ad hoc estimator is found to be a consistent estimator of θ only under a model that is not intuitively reasonable. This model requires the probability that an HU reports a victimization to be independent of the number of months in the year for which the HU contributes information. A modified version of this ad hoc estimator is consistent for θ under a more reasonable model. This model specifies that the probability that an HU contributing i months of information will report a victimization in those i months is $12\theta/[(12-i)\theta + i]$. Of all of the models discussed here this model fits the data best for each year. In addition, it fits the data from the HUs with no nonresponse and the data from the HUs with at least one nonresponse better than the other models. Although this model is not as easily interpretable as the Bernoulli and Markov models, it seems to more accurately describe the data.

The two BJS estimators are found to be consistent under two very similar models. These models require that HUs that have a type A noninterview or are out of sample for part of the year are victimized similarly to HUs that contribute a full year's information. In addition, HUs with a completed interview and a type B or C noninterview must have the same probability of victimization at that one completed interview as an HU that has contributed a full year's information has of being victimized in the entire year. Although this assumption seems unreasonable at first, we find in Chapter 5 that HUs with type B noninterviews do report more crimes than HUs with no nonresponse. Thus the BJS estimators may fortuitously account in part for information that is not missing at random.

After exploring these ad hoc estimators, we fit several models to the data for each of the three years 1973-1975 under the assumption that the missing data were missing at random. The first of these models was the homogeneous Bernoulli model under which every HU had the same probability of being victimized in any particular month. Victimization in any month was assumed to be independent of victimization in any other month and HUs were assumed to be victimized independently of one another. The fit of this model was poor for the 1973 and 1974 data. Although the fit was better for the 1975 data, the model overestimated the number of HUs contributing 12 months of information that were victimized for all three years.

We next fit a correlated Bernoulli model to the data but after examining the model in detail, dismissed it because of its poor fit to the data and also because of its unusual interpretation.

We then fit a Markov model to the three years of data and found results similar to those for the Bernoulli model. The fits for the 1973 and 1974 data were poor, but that for the 1975 data was acceptable. Since the Bernoulli model is a special case of the Markov model, we were able to use the results of the 1975 fit to find that the Markov model is an improvement over the Bernoulli model.

Comparing the estimates obtained from these models we noticed a slight tendency for the estimates to increase over time. We note in Chapter 6 that this may be due to rotation group biases and, in fact, the poor fit of the Bernoulli and Markov models to the 1973 and 1974 data may also be due to these biases. The estimates corresponding to the better fitting models are all in the .65 to .70 range and generally agree with the BJS estimates. In figures 13 and 14 we compared the estimated probability of victimization as a function of number of months of information contributed for the different models. We found that this function for the consistency model increases less as the number of months of information increases than for the Markov and Bernoulli models.

We had assumed through the first 4 chapters that the missing data were missing at random. If this assumption were not valid, then the poor fits of the Markov and Bernoulli models to the 1973 and 1974 data could be due to the HUs that contributed nonresponses. To determine whether or not this was the case, we then fit the Bernoulli, Markov, and consistency models to two subsets of the data. The first subset consisted of all the HUs that had no nonresponse in the year of interest. The second subset consisted of those that had at least one nonresponse in the year of interest. Fitting the models to these two subsets did not dramatically improve the fits but we did find that HUs with at least one nonresponse have lower cheery indicators than HUs with no nonresponse. In addition, we found some evidence that HUs with type B noninterviews report more victimizations than HUs with no nonresponse and HUs with type A noninterviews.

We then attempted to incorporate the nonresponse in our model by fitting an ordered category logit model with explanatory variables "months of information" and "number of noninterviews". By incorporating the HUs that contributed only nonresponse in the year of interest in different ways, we computed three estimates of θ for 1973 and 1975. Unfortunately, this model did not fit the 1974 data well. The largest of these three estimates for each of 1973 and 1975 was close to the consistency estimate with the other two being smaller.

In Chapter 6, we did some preliminary estimation of θ at the HH level using the 1975 data and found that the estimates were slightly higher than the corresponding estimates at the HU level. In addition, we discussed the problem of rotation group biases in the National Crime Survey and suggested several directions for further research.

APPENDIX A

QUESTIONNAIRE

This copy of the NCS questionnaire has been taken from U.S. Department of Justice(1978). Although the questionnaire has changed from time to time throughout the survey years, this is a representative form. The questions on the first two pages are used to collect background information on the HU and the household respondent. The household and screen questions may be found on the third page of the questionnaire. If the response to any of these questions is "yes", a crime incident report is filled out. The first page of the crime incident report may be found on the last page of Appendix A.

O.M.B. No. 41-R2661; Approval Expires June 30, 1977

FORM NCS-1 AND NCS-2
(8-15-75)

U.S. DEPARTMENT OF COMMERCE
BUREAU OF THE CENSUS
ACTING AS COLLECTING AGENT FOR THE
LAW ENFORCEMENT ASSISTANCE ADMINISTRATION
U.S. DEPARTMENT OF JUSTICE

NATIONAL CRIME SURVEY
NATIONAL SAMPLE
NCS-1 – BASIC SCREEN QUESTIONNAIRE
NCS-2 – CRIME INCIDENT REPORT

NOTICE – Your report to the Census Bureau is confidential by law (U.S. Code 42, Section 3761). All identifiable information will be used only by persons engaged in and for the purposes of the survey, and may not be disclosed or released to others for any purpose.

Sample (cc 4) JO ____ | Control number (cc 5): PSU | Segment | Ck | Serial

Household number (cc 2) | Land use (cc 9–11)

NCS 1 and 2

INTERVIEWER: Fill Sample and Control numbers, and items 1, 2, 4, and 9 at time of interview.

1. Interviewer identification
(010) Code | Name

2. Record of interview
(011) Line number of household respondent (cc 12) | Date completed

3. TYPE Z NONINTERVIEW
Interview not obtained for
Line number
(016) ____
(017) ____
(018) ____
(019) ____

NOTE: Fill NCS-7 Noninterview Record, for Types A, B, and C noninterviews.

Complete 14–21 for each line number listed.

4. Household status
(020)
1 ☐ Same household as last enumeration
2 ☐ Replacement household since last enumeration
3 ☐ Previous noninterview or not in sample before

5. Special place type code (cc 6c)
(021) ____

6. Tenure (cc 8)
(022)
1 ☐ Owned or being bought
2 ☐ Rented for cash
3 ☐ No cash rent

7. Type of living quarters (cc 15)
(023)
Housing unit
1 ☐ House, apartment, flat
2 ☐ HU in nontransient hotel, motel, etc.
3 ☐ HU – Permanent in transient hotel, motel, etc.
4 ☐ HU in rooming house
5 ☐ Mobile home or trailer
6 ☐ HU not specified above – *Describe*

OTHER Unit
7 ☐ Quarters not HU in rooming or boarding house
8 ☐ Unit not permanent in transient hotel, motel, etc.
9 ☐ Vacant tent site or trailer site
10 ☐ Not specified above – *Describe*

8. Number of housing units in structure (cc 26)
(024)
1 ☐ 1
2 ☐ 2
3 ☐ 3
4 ☐ 4
5 ☐ 5–9
6 ☐ 10 or more
7 ☐ Mobile home or trailer
8 ☐ Only OTHER units

ASK IN EACH HOUSEHOLD:
9. (Other than the . . . business) does anyone in this household operate a business from this address?
(025)
1 ☐ No
2 ☐ Yes – What kind of business is that?

INTERVIEWER: Enter unrecognizable businesses only

CENSUS USE ONLY

(026) **10. Family income (cc 27)**
1 ☐ Under $1,000
2 ☐ $1,000 to 1,999
3 ☐ 2,000 to 2,999
4 ☐ 3,000 to 3,999
5 ☐ 4,000 to 4,999
6 ☐ 5,000 to 5,999
7 ☐ 6,000 to 7,499
8 ☐ 7,500 to 9,999
9 ☐ 10,000 to 11,999
10 ☐ 12,000 to 14,999
11 ☐ 15,000 to 19,999
12 ☐ 20,000 to 24,999
13 ☐ 25,000 to 49,999
14 ☐ 50,000 and over

11a. Household members 12 years of age and OVER
(027) ____ Total number

b. Household members UNDER 12 years of age
(028) ____ Total number
0 ☐ None

12. Crime Incident Reports filled
(029) ____ Total number – *Fill item 31 on Control Card*
0 ☐ None

13a. Use of telephone (cc 25)
(030)
☐ Phone in unit (Yes in cc 25a)
Phone interview acceptable? (cc 25c or 25d)
1 ☐ Yes . . . . } *SKIP to next applicable item*
2 ☐ No – Refused number }
☐ Phone elsewhere (Yes in cc 25b)
Phone interview acceptable? (cc 25c or 25d)
3 ☐ Yes . . . . } *SKIP to next applicable item*
4 ☐ No – Refused number }
5 ☐ No phone (No in cc 25a and 25b)

13b. Proxy information – *Fill for all proxy interviews*

(1) Proxy interview obtained for line number ____
Proxy respondent name | Line number
Reason for proxy interview

(2) Proxy interview obtained for line number ____
Proxy respondent name | Line number
Reason for proxy interview

If more than 2 Proxy Interviews, continue in notes.

(031) | (032) | (033)

PERSONAL CHARACTERISTICS

14. NAME (of household respondent) KEYER – BEGIN NEW RECORD	15. TYPE OF INTERVIEW	16. LINE NO. (cc 12)	17. RELATIONSHIP TO HOUSEHOLD HEAD (cc 13b)	18. AGE LAST BIRTHDAY (cc 17)	19. MARITAL STATUS (cc 18)	20a. RACE (cc 19a)	20b. ORIGIN (cc 19b)	21. SEX (cc 20)	22. ARMED FORCES MEMBER (cc 21)	23. Education – highest grade (cc 22)	24. Education – complete that year? (cc 23)
Last First	(034) 1 ☐ Per – Self respondent 2 ☐ Tel. – Self-respondent 3 ☐ Per. – Proxy } Fill 13b on cover page 4 ☐ Tel. – Proxy } 5 ☐ NI – Fill 16–21	(035) Line No.	(036) 1 ☐ Head 2 ☐ Wife of head 3 ☐ Own child 4 ☐ Other relative 5 ☐ Non-relative	(037) Age	(038) 1 ☐ M. 2 ☐ Wd. 3 ☐ D. 4 ☐ Sep. 5 ☐ NM	(039) 1 ☐ W. 2 ☐ Neg. 3 ☐ Ot.	Origin	(040) 1 ☐ M 2 ☐ F	(041) 1 ☐ Yes 2 ☐ No	(042) Grade	(043) 1 ☐ Yes 2 ☐ No

CHECK ITEM A ➡ Look at item 4 on cover page. Is this the same household as last enumeration? *(Box 1 marked)*
☐ Yes – *SKIP* to *Check Item B* ☐ No

25a. Did you live in this house on April 1, 1970?
(044) 1 ☐ Yes – *SKIP to Check Item B* 2 ☐ No

b. Where did you live on April 1, 1970? (State, foreign country, U.S. possession, etc.)
State, etc. ______________ County ______________

c. Did you live inside the limits of a city, town, village, etc.?
(045) 1 ☐ No 2 ☐ Yes – *Name of city, town, village, etc.*
(046) ☐☐☐☐☐

(Ask males 18+ only)
d. Were you in the Armed Forces on April 1, 1970?
(047) 1 ☐ Yes 2 ☐ No

CHECK ITEM B ➡ Is this person 16 years old or older?
☐ No – *SKIP* to 29 ☐ Yes

26a. What were you doing most of LAST WEEK – (working, keeping house, going to school) or something else?
(048)
1 ☐ Working – *SKIP to 28a*
2 ☐ With a job but not at work
3 ☐ Looking for work
4 ☐ Keeping house
5 ☐ Going to school
6 ☐ Unable to work – *SKIP to 26d*
7 ☐ Retired
8 ☐ Other – *Specify*
(If Armed Forces, SKIP to 28a)

b. Did you do any work at all LAST WEEK, not counting work around the house? *(Note: If farm or business operator in HH, ask about unpaid work.)*
(049) 0 ☐ No Yes – How many hours? ______ – *SKIP to 28a*

c. Did you have a job or business from which you were temporarily absent or on layoff LAST WEEK?
(050) 1 ☐ No 2 ☐ Yes – Absent – *SKIP* to 28a
3 ☐ Yes – Layoff – *SKIP* to 27

26d. Have you been looking for work during the past 4 weeks?
(051) 1 ☐ Yes No – When did you last work?
2 ☐ Less than 5 years ago – *SKIP to 28a*
3 ☐ 5 or more years ago } *SKIP* to 29
4 ☐ Never worked }

27. Is there any reason why you could not take a job LAST WEEK?
(052) 1 ☐ No Yes – 2 ☐ Already had a job
3 ☐ Temporary illness
4 ☐ Going to school
5 ☐ Other – Specify

28a. For whom did you (last) work? *(Name of company, business, organization or other employer)*
(053) x ☐ Never worked – *SKIP* to 29

b. What kind of business or industry is this? *(E.g.: TV and radio mfg., retail shoe store, State Labor Department, farm)*
(054) ☐☐☐

c. Were you –
(055) 1 ☐ An employee of a PRIVATE company, business or individual for wages, salary or commissions?
2 ☐ A GOVERNMENT employee (Federal, State, county, or local)?
3 ☐ SELF-EMPLOYED in OWN business, professional practice or farm?
4 ☐ Working WITHOUT PAY in family business or farm?

d. What kind of work were you doing? *(E.g.: electrical engineer, stock clerk, typist, farmer, Armed Forces)*
(056) ☐☐☐

e. What were your most important activities or duties? *(E.g.: typing, keeping account books, selling cars, Armed Forces)*

Notes

FORM NCS-1 (6-15-78)

HOUSEHOLD SCREEN QUESTIONS

29. Now I'd like to ask some questions about crime. They refer only to the last 6 months – between _______ 1, 197___ and _______, 197___. During the last 6 months, did anyone break into or somehow illegally get into your (apartment/home), garage, or another building on your property?
☐ Yes – How many times? ______
☐ No

30. (Other than the incident(s) just mentioned) Did you find a door jimmied, a lock forced, or any other signs of an ATTEMPTED break in?
☐ Yes – How many times? ______
☐ No

31. Was anything at all stolen that is kept outside your home, or happened to be left out, such as a bicycle, a garden hose, or lawn furniture? (other than any incidents already mentioned)
☐ Yes – How many times? ______
☐ No

32. Did anyone take something belonging to you or to any member of this household, from a place where you or they were temporarily staying, such as a friend's or relative's home, a hotel or motel, or a vacation home?
☐ Yes – How many times? ______
☐ No

33. What was the total number of motor vehicles (cars, trucks, etc.) owned by you or any other member of this household during the last 6 months?
(057)
0 ☐ None – *SKIP to 36*
1 ☐ 1
2 ☐ 2
3 ☐ 3
4 ☐ 4 or more

34. Did anyone steal, TRY to steal, or use (it/any of them) without permission?
☐ Yes – How many times? ______
☐ No

35. Did anyone steal or TRY to steal parts attached to (it/any of them), such as a battery, hubcaps, tape-deck, etc.?
☐ Yes – How many times? ______
☐ No

INDIVIDUAL SCREEN QUESTIONS

36. The following questions refer only to things that happened to YOU during the last 6 months – between _______ 1, 197___ and _______, 197___. Did you have your (pocket picked/purse snatched)?
☐ Yes – How many times? ______
☐ No

37. Did anyone take something (else) directly from you by using force, such as by a stickup, mugging or threat?
☐ Yes – How many times? ______
☐ No

38. Did anyone TRY to rob you by using force or threatening to harm you? (other than any incidents already mentioned)
☐ Yes – How many times? ______
☐ No

39. Did anyone beat you up, attack you or hit you with something, such as a rock or bottle? (other than any incidents already mentioned)
☐ Yes – How many times? ______
☐ No

40. Were you knifed, shot at, or attacked with some other weapon by anyone at all? (other than any incidents already mentioned)
☐ Yes – How many times? ______
☐ No

41. Did anyone THREATEN to beat you up or THREATEN you with a knife, gun, or some other weapon, NOT including telephone threats? (other than any incidents already mentioned)
☐ Yes – How many times? ______
☐ No

42. Did anyone TRY to attack you in some other way? (other than any incidents already mentioned)
☐ Yes – How many times? ______
☐ No

43. During the last 6 months, did anyone steal things that belonged to you from inside ANY car or truck, such as packages or clothing?
☐ Yes – How many times? ______
☐ No

44. Was anything stolen from you while you were away from home, for instance at work, in a theater or restaurant, or while traveling?
☐ Yes – How many times? ______
☐ No

45. (Other than any incidents you've already mentioned) was anything (else) at all stolen from you during the last 6 months?
☐ Yes – How many times? ______
☐ No

46. Did you find any evidence that someone ATTEMPTED to steal something that belonged to you? (other than any incidents already mentioned)
☐ Yes – How many times? ______
☐ No

47. Did you call the police during the last 6 months to report something that happened to YOU which you thought was a crime? (Do not count any calls made to the police concerning the incidents you ~~have just told me about.~~)
☐ No – *SKIP to 48*
☐ Yes – What happened?
______
______
______
(058) ☐☐ ☐☐ ☐☐

CHECK ITEM C ➜ Look at 47. Was HH member 12+ attacked or threatened, or was something stolen or an attempt made to steal something that belonged to him?
☐ Yes – How many times? ______
☐ No

48. Did anything happen to YOU during the last 6 months which you thought was a crime, but did NOT report to the police? (other than any incidents already mentioned)
☐ No – *SKIP to Check Item E*
☐ Yes – What happened?
______
______
______
(059) ☐☐ ☐☐ ☐☐

CHECK ITEM D ➜ Look at 48. Was HH member 12+ attacked or threatened, or was something stolen or an attempt made to steal something that belonged to him?
☐ Yes – How many times? ______
☐ No

CHECK ITEM E ➜ Do any of the screen questions contain any entries for "How many times?"
☐ No – *Interview next HH member. End interview if last respondent, and fill item 12 on cover page.*
☐ Yes – *Fill Crime Incident Reports.*

FORM NCS-1 (6-15-75)

O.M.B. No. 41-R2661; Approval Expires June 30, 1977

KEYER – BEGIN NEW RECORD

Notes

NOTICE – Your report to the Census Bureau is confidential by law (U.S. Code 42, Section 3761). All identifiable information will be used only by persons engaged in and for the purposes of the survey, and may not be disclosed or released to others for any purpose.

FORM NCS-2
(6-15-76)

U.S. DEPARTMENT OF COMMERCE
BUREAU OF THE CENSUS
ACTING AS COLLECTING AGENT FOR THE
LAW ENFORCEMENT ASSISTANCE ADMINISTRATION
U.S. DEPARTMENT OF JUSTICE

CRIME INCIDENT REPORT
NATIONAL CRIME SURVEY – NATIONAL SAMPLE

NCS 2 INCIDENT REPORT

(101) Line number

(102) Screen question number

(103) Incident number

1a. You said that during the last 6 months – *(Refer to appropriate screen question for description of crime).*
In what month (did this/did the first) incident happen? *(Show flashcard if necessary. Encourage respondent to give exact month.)*

(104) ________ Month (01–12) | Year 197____

(105) **CHECK ITEM A** Is this incident report for a series of crimes?
1 ☐ No – *SKIP to 2*
2 ☐ Yes – *(Note: series must have 3 or more similar incidents which respondent can't recall separately)*

b. In what month(s) did these incidents take place? *(Mark all that apply)*
(106) *
1 ☐ Spring (March, April, May)
2 ☐ Summer (June, July, August)
3 ☐ Fall (September, October, November)
4 ☐ Winter (December, January, February)

c. How many incidents were involved in this series?
(107)
1 ☐ Three or four
2 ☐ Five to ten
3 ☐ Eleven or more
4 ☐ Don't know

INTERVIEWER: If this report is for a series, read the following statement.
(The following questions refer only to the most recent incident.)

2. About what time did (this/the most recent) incident happen?
(108)
1 ☐ Don't know
2 ☐ During the day (6 a.m. to 6 p.m.)
At night (6 p.m. to 6 a.m.)
3 ☐ 6 p.m. to midnight
4 ☐ Midnight to 6 a.m.
5 ☐ Don't know

3a. In what State and county did this incident occur?
☐ Outside U.S. – *END INCIDENT REPORT*
State ____________ County ____________

b. Did it happen INSIDE THE LIMITS of a city, town, village, etc.?
(110)
1 ☐ No
2 ☐ Yes – *Enter name of city, town, etc.* ↴
(111) ☐☐☐☐☐☐ ____________

4. Where did this incident take place?
(112)
1 ☐ At or in own dwelling, in garage or other building on property *(Includes break-in or attempted break-in)* — *SKIP to 6a*
2 ☐ At or in a vacation home, hotel motel — *SKIP to 6a*
3 ☐ Inside commercial building such as store, restaurant, bank, gas station, public conveyance or station — *ASK 5a*
4 ☐ Inside office, factory, or warehouse — *ASK 5a*
5 ☐ Near own home; yard, sidewalk, driveway, carport, apartment hall *(Does not include break-in or attempted break-in)* — *SKIP to Check Item B*
6 ☐ On the street, in a park, field, playground, school grounds or parking lot — *SKIP to Check Item B*
7 ☐ Inside school — *SKIP to Check Item B*
8 ☐ Other – *Specify* ↴ — *SKIP to Check Item B*
____

5a. Were you a customer, employee, or owner?
(113)
1 ☐ Customer
2 ☐ Employee
3 ☐ Owner
4 ☐ Other – *Specify* ____________

b. Did the person(s) steal or TRY to steal anything belonging to the store, restaurant, office, factory, etc.?
(114)
1 ☐ Yes
2 ☐ No
3 ☐ Don't know
} *SKIP to Check Item B*

6a. Did the offender(s) live there or have a right to be there, such as a guest or a workman?
(115)
1 ☐ Yes – *SKIP to Check Item B*
2 ☐ No
3 ☐ Don't know

b. Did the offender(s) actually get in or just TRY to get in the building?
(116)
1 ☐ Actually got in
2 ☐ Just tried to get in
3 ☐ Don't know

c. Was there any evidence, such as a broken lock or broken window, that the offender(s) (forced his way in/TRIED to force his way in) the building?
(117) *
1 ☐ No
Yes – **What was the evidence? Anything else?** *(Mark all that apply)*
2 ☐ Broken lock or window
3 ☐ Forced door or window
4 ☐ Slashed screen
5 ☐ Other – *Specify* ↴
____
} *SKIP to Check Item B*

d. How did the offender(s) (get in/try to get in)?
(118)
1 ☐ Through unlocked door or window
2 ☐ Had key
3 ☐ Don't know
4 ☐ Other – *Specify* ____________

(119) **CHECK ITEM B** Was respondent or any other member of this household present when this incident occurred? *(If not sure, ASK)*
1 ☐ No – *SKIP to 13a*
2 ☐ Yes

7a. Did the person(s) have a weapon such as a gun or knife, or something he was using as a weapon, such as a bottle, or wrench?
(120) *
1 ☐ No
2 ☐ Don't know
Yes – **What was the weapon? Anything else?** *(Mark all that apply)*
3 ☐ Gun
4 ☐ Knife
5 ☐ Other – *Specify* ____________

b. Did the person(s) hit you, knock you down, or actually attack you in any way?
(121)
1 ☐ Yes – *SKIP to 7f*
2 ☐ No

c. Did the person(s) threaten you with harm in any way?
(122)
1 ☐ No – *SKIP to 7e*
2 ☐ Yes

APPENDIX B

COMPUTER PROGRAMS

Program 1

This program collects the statistics from the data tapes that are needed to test the fit of the Markov model. It is representative of the programs that were used to extract statistics from the data files.

```
CCCCCCCCCCCCCCCCCCCCCCCCCCCCCCCCCCCCCCCCCCCCCCCCCCCCCCCCCCCCCC
C                                                             C
C   THIS PROGRAM CALCULATES THE TABLE OF HUS BY PATTERNS OF   C
C   MISSINGNESS.  THERE ARE 4 BASIC POSSIBILITIES:            C
C        1 IN FOR I MONTHS ,OUT FOR 12-I   I GE 1             C
C              THESE ARE STORED IN FIRIN                      C
C        2 IN FOR I MONTHS, OUT FOR 6, IN FOR 6-I  0<I<6      C
C              THESE ARE STORED IN MIDO                       C
C        3 OUT FOR I MONTHS, IN FOR 12-I, 0<I<12              C
C              THESE ARE STORED IN LASIN                      C
C        4 OUT FOR I MONTHS, IN FOR 6, AND OUT 6-I, 0<I<6     C
C              THESE ARE STORED IN MIDIN                      C
C                                                             C
C   THIS IS DONE BY CREATING A 12 BY 3 WORKING ARRAY (MONTH   C
C   BY YEAR) AND PUTTING THE APPROPRIATE CODE INTO EACH       C
C   ELEMENT.  THE CODES ARE:                                  C
C                                                             C
C            -2 - NOT IN SAMPLE                               C
C            -1 - NON-INTERVIEW                               C
C             0 - NO VICTIMIZATION                            C
C             1 - CRIME                                       C
C   FROM THE WORKING ARRAY IT IS THEN DETERMINED, FOR EACH    C
C   TO WHICH OF THE ABOVE GROUPS THE HU BELONGS AND IT IS     C
C   ADDED TO THE APPROPRIATE GROUP                            C
C                                                             C
C   I O      :                                                C
C                                                             C
C            20 - THE HH FILE                                 C
C            21 - THE INCIDENT FILE                           C
C            22 - OUTPUT                                      C
C                                                             C
C   SUBROUTINES :                                             C
C                                                             C
C      FMTHIS - CALCULATES THE FIRST COVERED MONTH OF DATA    C
```

```
C      INTER - FILLS THE WORKING ARRAY WITH 0 FOR EACH         C
C              INTERVIEW.                                      C
C      CRIME - FILLS THE WORKING ARRAY WITH 1 FOR EACH CRIME   C
C      COUNT - CALCULATES THE TABLES OF COUNTS                 C
C      OUTPUT - WRITES THE OUTPUT ON DISK.                     C
C                                                              C
CCCCCCCCCCCCCCCCCCCCCCCCCCCCCCCCCCCCCCCCCCCCCCCCCCCCCCCCCCCCCCCC

      LOGICAL INEOF,ERROR
      INTEGER HHDATE(2),INDATE(2),HHID(10),INID(10),CURID(10),INTYPE,
     +WORKA(12,3),FIRIN(12,2,3),MIDO(5,2,3),LASIN(11,2,3),MIDIN(5,2,3)
      DATA INEOF/.FALSE./,WORKA/36*-2/
      OPEN(UNIT=20,STATUS='OLD',FILE='DATA.HH')
      OPEN(UNIT=21,STATUS='OLD',FILE='DATA.IN')
      OPEN(UNIT=22,STATUS='NEW',CARRIAGECONTROL='LIST',
     _FILE='P1.OUT')

C READ THE ID, INTERVIEW DATE, AND TYPE OF INTERVIEW FROM
C THE FIRST HU RECORD
      READ(20,1000)HHID,HHDATE(1),HHDATE(2),INTYPE

C CALCULATE FROM THE INTERVIEW DATE THE FIRST MONTH COVERED BY
C DATA FOR THIS INTERVIEW
      CALL FMTHIS ( HHID,HHDATE,FSAM,ERROR)

C READ THE ID, AND INCIDENT DATE FROM THE FIRST INCIDENT RECORD
      READ(21,1100)INID,INDATE(1),INDATE(2)

C SET CURRENT ID EQUAL TO THE HU ID
      DO 33 KK=1,10
      CURID(KK)=HHID(KK)
33    CONTINUE

C SEE IF THIS HU RECORD IS FROM THE SAME HU AS THE PREVIOUS HU RECORD
100   DO 133 KK=1,10
      IF ( CURID(KK) .NE. HHID(KK) )GO TO 200
133   CONTINUE

C PUT ZEROS IN THE WORKING ARRAY FOR EACH MONTH THAT WAS COVERED BY AN
C   INTERVIEW
150   CALL INTER(WORKA,HHDATE,INTYPE)

C READ THE ID, INTERVIEW DATE AND INTERVIEW TYPE FROM THE NEXT HU RECORD
      READ(20,1000,END=600)HHID,HHDATE(1),HHDATE(2),INTYPE

      GO TO 100


C UNLESS THERE ARE NO MORE INCIDENTS, SEE IF THE INCIDENT BELONGS TO
C   THE CURRENT HU
200   IF ( INEOF ) GO TO 300
         DO 233 KK=1,10
         IF ( CURID(KK) .NE. INID(KK) ) GO TO 300
233      CONTINUE

C IF THE INCIDENT BELONGS TO THE CURRENT HU, PUT A 1 IN THE WORKING
C   ARRAY AT THE MONTH OF THE INCIDENT, GET ANOTHER INCIDENT AND
C   SEE IF IT ALSO BELONGS TO THE CURRENT HU
```

```
      CALL CRIME(WORKA,INDATE)
      READ(21,1100,END=500)INID,INDATE(1),INDATE(2)
      GO TO 200

C ALL OF THE INFORMATION FOR THE HU HAS BEEN INCORPORATED INTO THE
C   WORKING ARRAY.  ADD THE HU INTO THE APPROPRIATE TABLE
300   CALL COUNT(WORKA,FIRIN,MIDO,LASIN,MIDIN)

C REINITIALIZE EVERYTHING
      DO 303 KK=1,10
      CURID(KK)=HHID(KK)
303   CONTINUE
      DO 350 I = 1,12
         DO 325 J=1,3
            WORKA(I,J)=-2
 325     CONTINUE
 350  CONTINUE
      CALL FMTHIS ( HHID,HHDATE,FSAM,ERROR)

C START A NEW WORKING ARRAY
      GO TO 150

C END OF INCIDENT FILE
500   INEOF = .TRUE.
      GO TO 300

C THIS IS THE LAST HU. IF THERE ARE STILL SOME INCIDENT RECORDS LEFT
C   THEY MUST ALL BELONG TO THE LAST HU.  READ THEM ALL AND ENTER
C   THE INCIDENTS INTO THE WORKING ARRAY
600   IF ( INEOF ) GO TO 800
700   CALL CRIME(WORKA,INDATE)
      READ (21,1100,END=800)INID,INDATE(1),INDATE(2)
      GO TO 700


C ALL OF THE INFO FOR THE LAST HU IS IN THE WORKING ARRAY.  ADD
C   THIS LAST HU TO THE APPROPRIATE TABLE
800   CALL COUNT(WORKA,FIRIN,MIDO,LASIN,MIDIN)


C OUTPUT THE RESULTS
      CALL OUTPUT(FIRIN,MIDO,LASIN,MIDIN)
1000  FORMAT(1X,10I1,3X,2I2,5X,I1)
1100  FORMAT(1X,10I1,9X,2I2)

      STOP
      END



C THE SUBROUTINES

      SUBROUTINE FMTHIS ( ID,DATE,IFIRST,ERR)
C CALCULATES THE FIRST MONTH COVERED BY THE INTERVIEW
C TAKING PLACE ON DATE

      INTEGER ID(10),DATE(2),IFIRST,ADD
      LOGICAL ERR
```

```
      ERR = .FALSE.
      IF ( DATE(1) .LT. 72 .OR. DATE(1) .GT. 76 ) GO TO 400
      IF ( DATE(2) .LT. 1 .OR. DATE(2) .GT. 12 ) GO TO 400
      ISAM = ID(1)
      IF ( ISAM .LE. 0 .OR. ISAM .GE. 5 ) GO TO 400
      IPAN = ID(5)
      IROT = ID(6)
      ADD = 6
      IF (ISAM .EQ. 2 ) ADD = 12
      IF ( ISAM .EQ. 3 .OR. ISAM .EQ. 4 ) ADD = 24 + (IROT - 1)*6
      IFIRST =  ADD + IPAN - 6
      RETURN
 400  ERR = .TRUE.
9999  FORMAT(I,1X,5(I3,1X))
      RETURN
      END


      SUBROUTINE INTER(ARRAY,DATE,TYPE)
C PUTS A 0 IN THE WORKING ARRAY FOR EACH MONTH THAT WAS COVERED BY
C   THE INTERVIEW THAT TOOK PLACE ON DATE

      INTEGER ARRAY(36),DATE(2),TYPE
      INTEGER YR,MTH,TEST,START,STOP,FILL
      YR=DATE(1)
      MTH=DATE(2)
      TEST=YR*100+MTH
      IF ( TEST .LE. 7301 .OR. TEST .GE. 7607 ) RETURN
      MTH=MTH-6
      IF ( MTH .GT. 0 ) GO TO 100
         MTH = MTH +12
         YR=YR-1
100   START = (YR-73)*12 + MTH
      STOP = START+5
      START = MAX0(1,START)
      STOP = MIN0(36,STOP)
      FILL = -1
      IF ( TYPE .EQ. 4) FILL = 0
      DO 200 I = START,STOP
         ARRAY(I) = FILL
 200  CONTINUE
      RETURN
 9999 FORMAT(10I5)
      END


      SUBROUTINE CRIME(ARRAY,DATE)
C PUTS A 1 IN THE WORKING ARRAY AT THE MONTH DURING WHICH
C THE CRIME OCCURRED

      INTEGER ARRAY(36),DATE(2)
      INTEGER TEST,CR
      TEST = DATE(1)*100 + DATE(2)
      IF ( TEST .LT. 7301 .OR. TEST .GT.7512 ) RETURN
      CR = (DATE(1)-73)*12 +DATE(2)
      ARRAY(CR) = 1
      RETURN
      END
```

```
      SUBROUTINE COUNT(ARRAY,FIRIN,MIDO,LASIN,MIDIN)
C FOR EACH OF THE 3 YEARS, DECIDES TO WHICH OF THE 4 GROUPS THE
C HU BELONGS AND ADDS IT TO THE APPROPRIATE MATRIX

      INTEGER ARRAY(12,3),FIRIN(12,2,3),MIDO(5,2,3),LASIN(11,2,3)
     *,MIDIN(5,2,3)
      INTEGER IN(3),OUT(3),CODE(3)

      DO 100 IYR=1,3
         IST=1
         ICR=0
         IMO=1
         DO 101 II=1,3
            IN(II)=0
            OUT(II)=0
101      CONTINUE
         IF(ARRAY(1,IYR).LT.0)GO TO 500
         IF(ARRAY(1,IYR).GT.0)ICR=1

C  INTERVIEW MONTHS
150      IN(IST)=IN(IST)+1
         IMO=IMO+1
         IF(IMO.GT.12)GO TO 800
200      IF(ARRAY(IMO,IYR).LT.0)GO TO 480
         IF(ARRAY(IMO,IYR).GT.0)ICR=1
         IN(IST)=IN(IST)+1
         IMO=IMO+1
         IF(IMO.GT.12)GO TO 800
         GO TO 200

C  MISSING MONTHS
480      IST=IST+1
500      OUT(IST)=OUT(IST)+1
         IMO=IMO+1
         IF(IMO.GT.12)GO TO 800
550      IF(ARRAY(IMO,IYR).GE.0)GO TO 600
         OUT(IST)=OUT(IST)+1
         IMO=IMO+1
         IF(IMO.GT.12)GO TO 800
         GO TO 550

C  BACK TO INTERVIEW MONTHS
600      IF(ARRAY(IMO,IYR).GT.0)ICR=1
         IST=IST+1
         GO TO 150

C  END OF YEAR
800      IF(IN(1).GT.0.AND.OUT(1).GT.0)WRITE(5,900)
900      FORMAT(' ERROR')
         INJAN=1
         IF(OUT(1).GT.0)INJAN=0
         DO 850 KK=1,3
            CODE(KK)=IN(KK)+OUT(KK)
850      CONTINUE

         IF(INJAN.EQ.0)GO TO 875
```

```
           IF(CODE(3).GT.0)GO TO 860

C  ADD HU TO FIRIN
           FIRIN(CODE(1),1,IYR)= FIRIN(CODE(1),1,IYR)+1
           IF(ICR.EQ.1)FIRIN(CODE(1),2,IYR)=FIRIN(CODE(1),2,IYR)+1
           GO TO 100

C  ADD TO MIDO
860        MIDO(CODE(1),1,IYR)=MIDO(CODE(1),1,IYR)+1
           IF(ICR.EQ.1)MIDO(CODE(1),2,IYR)=MIDO(CODE(1),2,IYR)+1
           GO TO 100

875       IF(CODE(1).EQ.12)GO TO 100
          IF(CODE(3).NE.0)GO TO 880

C  ADD TO LASIN
          LASIN(CODE(1),1,IYR)=LASIN(CODE(1),1,IYR)+1
          IF(ICR.EQ.1)LASIN(CODE(1),2,IYR)=LASIN(CODE(1),2,IYR)+1
          GO TO 100

C ADD TO MIDIN
880       MIDIN(CODE(1),1,IYR)=MIDIN(CODE(1),1,IYR)+1
          IF(ICR.EQ.1)MIDIN(CODE(1),2,IYR)=MIDIN(CODE(1),2,IYR)+1

100     CONTINUE

        RETURN
        END


        SUBROUTINE OUTPUT(FIRIN,MIDO,LASIN,MIDIN)
C OUTPUTS THE RESULTS

        INTEGER L(3),FIRIN(12,2,3),MIDO(5,2,3),LASIN(11,2,3),MIDIN(5,2,3)
        DO 1 K=1,3
           L(K)=K+2
1       CONTINUE

        DO 500 I=1,3
        WRITE(22,50)L(I)
50      FORMAT(//' 197',I1)
        WRITE(22,60)
60      FORMAT('  I ',' I IN, REST OUT'/5X,'TOTAL','  VIC')
        WRITE(22,65)((J,FIRIN(J,1,I),FIRIN(J,2,I)),J=1,12)
65      FORMAT(1X,I4,2I5)
        WRITE(22,70)
70      FORMAT(/'  I ',' I IN,6 OUT,6-I IN'/5X,'TOTAL','  VIC')
        WRITE(22,65)((J,MIDO(J,1,I),MIDO(J,2,I)),J=1,5)
        WRITE(22,80)
80      FORMAT(/'  I ',' I OUT, 12-I IN'/5X,'TOTAL','  VIC')
        WRITE(22,65)((J,LASIN(J,1,I),LASIN(J,2,I)),J=1,11)
        WRITE(22,90)
90      FORMAT(/'  I  I OUT, 6 IN, 6-I OUT'/5X,'TOTAL  VIC')
        WRITE(22,65)((J,MIDIN(J,1,I),MIDIN(J,2,I)),J=1,5)

500     CONTINUE
        RETURN
        END
```

Program 2

This program computes the estimated moments of the homogeneous Bernoulli estimator.

```
CCCCCCCCCCCCCCCCCCCCCCCCCCCCCCCCCCCCCCCCCCCCCCCCCCCCCCCCCCCCCCCCCCCCCCC
C  THIS PROGRAM COMPUTES THE ESTIMATED MOMENTS OF THE HOMOGENEOUS    C
C  BERNOULLI ESTIMATOR.  THE USER MUST ENTER R, THE MOMENT TO BE     C
C  ESTIMATED                                                          C
C                                                                     C
C  UNIT 10 IS USED FOR OUTPUT                                         C
CCCCCCCCCCCCCCCCCCCCCCCCCCCCCCCCCCCCCCCCCCCCCCCCCCCCCCCCCCCCCCCCCCCCCCC

      DOUBLE PRECISION ANS,TEMP(24,25),COEF,Q,CTEMP(24),S1,S2,PART(24)
      INTEGER R,T
      OPEN(UNIT=10,DEVICE='DSK',FILE='P2.OUT')

C  INITIALIZE
      ANS=0.D0
      T=8182
      Q=7819.D0/8182.D0

C READ R, THE MOMENT TO BE ESTIMATED, R BETWEEN 1 AND 24, INCLUSIVE
      READ(5,20)R
20    FORMAT(I2)

C FOR EACH VALUE OF J
      DO 50 J=1,R
C CALCULATE THE BRACKETED PART OF EXPRESSION (4-7)
         DO 45 K=J,0,-1
            TEMP(J,K+1)=CHOOSE(J,K)*((-1.D0)**K)*(DFLOAT(J-K)**R)
            CTEMP(J)=CTEMP(J)+TEMP(J,K+1)
45       CONTINUE
      KEND=J+1
      WRITE(10,22)(TEMP(J,K),K=1,KEND)
22    FORMAT(F40.1)

C  CALCULATE THE FIRST AND SECOND TERMS OF EXPRESSION (4-9)
      S1=0.D0
      S2=0.D0
      DO 47 K=1,J
         S1=S1+DLOG(DFLOAT(T+1-K))
         S2=S2+DLOG(DFLOAT(K))
47    CONTINUE

C  ADD ALL TERMS OF EXPRESSION (4-9) FOR THIS VALUE OF J
      PART(J)=DEXP(S1-S2+DLOG(CTEMP(J))+DFLOAT(J)*DLOG(Q)-
     *DFLOAT(R)*DLOG(DFLOAT(T)))
50    CONTINUE

C WRITE OUT INTERMEDIATE CALCULATIONS
      WRITE(10,24)(CTEMP(J),J=1,R)
24    FORMAT(24F40.1)
      WRITE(10,25)(PART(J),J=1,R)
25    FORMAT(25F20.15)
```

```
C SUM EXPRESSION (4-9) OVER J
      DO 70 J=1,R
         ANS=ANS+PART(J)
70    CONTINUE

C WRITE RESULT
      WRITE(10,26)ANS
26    FORMAT(F25.10)
      STOP
      END

      DOUBLE PRECISION FUNCTION CHOOSE(A,B)
C THIS SUBROUTINE CALCULATES A CHOOSE B
      INTEGER A,B,BB
      DOUBLE PRECISION C
      C=1.D0
      IF(B.GT.A)GO TO 900
      IF((B.EQ.A).OR.(B.EQ.0))GO TO 910
      BB=B
      IF(B.LT.0.5*A)BB=A-B
      DO 300 I=1,A-BB
         C=C*((DFLOAT(BB+I))/DFLOAT(I))
300   CONTINUE
      CHOOSE=C
      RETURN

900   WRITE(5,21)
21    FORMAT(' ERROR IN BINOMIAL COEFFICIENT')
      CHOOSE=0.D0
      RETURN

910   CHOOSE=C
      RETURN
      END
```

Program 3

This program maximizes the likelihood for the correlated Bernoulli model.

```
CCCCCCCCCCCCCCCCCCCCCCCCCCCCCCCCCCCCCCCCCCCCCCCCCCCCCCCCCCCCCCCCCCCCCCC
C THIS PROGRAM CALCULATES THE MAX OF THE LIKELIHOOD FOR THE CORRELATED C
C BERNOULLI MODEL USING ZXMIN.                                         C
C VI=L-1, MI=K-1, J=B, P=X(1), RHO=X(2).                               C
C                                                                      C
C THE INPUT IS A MATRIX OF THE HUS CLASSIFIED BY MI AND VI AND IS READ C
C FROM UNIT 12                                                         C
CCCCCCCCCCCCCCCCCCCCCCCCCCCCCCCCCCCCCCCCCCCCCCCCCCCCCCCCCCCCCCCCCCCCCCC

        SUBROUTINE FUNCT(NNN,X,F)
C THIS SUBROUTINE COMPUTES -1 TIMES THE LOG LIKELIHOOD
C INITIALIZE
        INTEGER RES(13,13),B,C,D,BF,DF,LIM,LIM1,LIM2,LIM3
        DOUBLE PRECISION A,A1,A2,A3,ANS,X(2),F
        COMMON RES
        WRITE(13,*)X

C IF P AND RHO ARE NOT IN THE PARAMETER SPACE, SET F TO A LARGE NUMBER
        IF(X(1).GT.0.AND.X(1).LT.1.AND.X(2).GE.0.AND.X(2).LE.1)
     *    GO TO 9999
           F=9999999.99*10000000.
           RETURN
9999    CONTINUE

           ANS=0.D0
C LOOP ON ELEMENTS OF INPUT MATRIX
           DO 26 K=1,12
           LIM=14-K
           DO 36 L=1,LIM
              IF(RES(K,L).EQ.0) GO TO 36
              A2=0.D0
              A3=0.D0

C CALCULATE FIRST PART OF LIKELIHOOD
              IF(L.EQ.1) A1=((1.D0-X(1))**12)*(1.D0-X(2))+X(2)*
     *          (1.D0-X(1))
              IF(L.EQ.13) A1=(X(1)**12)*(1.D0-X(2))+X(2)*X(1)
              IF((L.GT.1).AND.(L.LT.13))A1=(X(1)**(L-1))*((1.D0-X(1))**
     2            (13-L))*(1.D0-X(2))

C CALCULATE SECOND PART OF LIKELIHOOD
              IF(K.LE.2)GO TO 24
C            CALCULATE M!
              LIM1=K-1
              MF=1
              DO 20 N=1,LIM1
              MF=MF*N
20            CONTINUE

              LIM2=K-2
              DO 23 B=1,LIM2
C            CALCULATE B!
```

```
              BF=1
              DO 21 C=1,B
              BF=BF*C
21            CONTINUE

C            CALCULATE (M-B)!
              LIM3=K-1-B
              DF=1
              DO 22 D=1,LIM3
              DF=DF*D
22            CONTINUE

              A2=A2+(DFLOAT(MF)/(DFLOAT(BF)*DFLOAT(DF)))*(X(1)**(L-1+B))
    1               *((1.D0-X(1))**(13-B-L))*(1.D0-X(2))

23            CONTINUE

C CALCULATE THIRD PART OF LIKELIHOOD
24          IF(K.EQ.1) GO TO 25
            IF(L.EQ.(14-K)) A3=(X(1)**12)*(1.D0-X(2))+X(2)*X(1)
            IF(L.LT.(14-K)) A3=(X(1)**(L+K-2))*((1.D0-X(1))**(14-L-K))*
    *       (1.D0-X(2))

C COMBINE PARTS
25          A=(DLOG(A1+A2+A3))*DFLOAT(RES(K,L))
            ANS=ANS+A
        F=-1.D0*ANS

36      CONTINUE
26      CONTINUE

        RETURN
        END


C MAIN PROGRAM
        EXTERNAL FUNCT
        INTEGER RES(13,13)
        DOUBLE PRECISION X(2),H(3),G(2),W(6)
        COMMON RES
        OPEN(UNIT=12,DEVICE='DSK',FILE='P3.IN')
        OPEN(UNIT=13,DEVICE='DSK',FILE='P3.OUT')

C SET NUMBER OF PARAMETERS, NUMBER OF SIGNIFICANT DIGITS DESIRED,
C MAXIMIM NUMBER OF FUNCTION EVALUATIONS
        N=2
        NSIG=3
        MAXFN=1000

C INITIALIZE HESSIAN TO IDENTITY AND GIVE INITIAL PARAMETER ESTIMATES
        IOPT=0
        X(1)=.071620D0
        X(2)=.417079D0

C READ INPUT MATRIX
        READ(12,77)RES
77      FORMAT(13I3)
```

```
C CALL ZXMIN TO MINIMIZE -1 TIMES THE LOG LIKELIHOOD
        CALL ZXMIN(FUNCT,N,NSIG,MAXFN,IOPT,X,H,G,F,W,IER)

C WRITE THE OUTPUT
        WRITE(13,1)X
1       FORMAT(1X,'P=',F10.6,1X,'RHO=',F10.6)
        WRITE(13,2)H
2       FORMAT(1X,'HESSIAN IS',3G)
        WRITE(13,3)(W(I),I=1,3)
3       FORMAT(1X,'NORM OF GRAD',F10.6/1X,'STEPS',F10.6/1X,'SIG.
     *  DIGS',F10.6)
        WRITE(13,4)IER
4       FORMAT(1X,I4)
        WRITE(13,5)F,G
5       FORMAT(1X,'F=',G,3X,'GRAD =',2G)
        STOP
        END
```

Program 4

This program maximizes the Markov model likelihood.

```
CCCCCCCCCCCCCCCCCCCCCCCCCCCCCCCCCCCCCCCCCCCCCCCCCCCCCCCCCCCCCCCC
C                                                              C
C   THIS PROGRAM MAXIMIZES THE MARKOV LIKELIHOOD USING THE IMSL C
C   SUBROUTINE ZXMIN.  -1 TIMES THE LOG LIKELIHOOD IS COMPUTED  C
C   BY THE SUBROUTINE FUNCT AND IS MINIMIZED BY ZXMIN           C
C                                                              C
C   I O:                                                       C
C        10 - INPUT                                            C
C        11 - OUTPUT                                           C
C                                                              C
CCCCCCCCCCCCCCCCCCCCCCCCCCCCCCCCCCCCCCCCCCCCCCCCCCCCCCCCCCCCCCCC


        SUBROUTINE FUNCT(NN,Y,F)
C CALCULATES -1 TIMES THE LOG LIKELIHOOD

        DOUBLE PRECISION Y(3),F
        DOUBLE PRECISION INIT(2),P(2,2),LOGLIK,X(3),PARLIK
        DOUBLE PRECISION TEMP
        INTEGER POW,REM,PLACE,MISS(8,11),POW2
        INTEGER POS1(6,64),POS2(11,4096)

C RENAME THE Y VECTOR TO X
        DO 9 I=1,3
           X(I)=Y(I)
9       CONTINUE

C MAKE SURE THE PARAMETERS ARE IN THE PARAMETER SPACE
        IF(X(1).LE.0.D0.OR.X(1).GE.1.D0)GO TO 666
        IF(X(2).LE.0.D0.OR.X(2).GE.1.D0)GO TO 666
        IF(X(3).LE.0.D0.OR.X(3).GE.1.D0)GO TO 666

C RENAME VARIABLES SO THE PROGRAM IS EASIER TO UNDERSTAND!
        INIT(1)=X(1)
        P(1,1)=X(2)
        P(2,1)=X(3)
        P(1,2)=1.D0-P(1,1)
        P(2,2)=1.D0-P(2,1)
        INIT(2)=1.D0-INIT(1)

C ADD ALL OF THE CONTRIBUTIONS OF THE OBSERVED TRANSITIONS INTO
C THE LOG LIKELIHOOD
        LOGLIK=DFLOAT(N0)*DLOG(INIT(1))+DFLOAT(N1)*DLOG(INIT(2))+
        1DFLOAT(N00)*DLOG(P(1,1))+DFLOAT(N01)*DLOG(P(1,2))+
        2DFLOAT(N10)*DLOG(P(2,1))+DFLOAT(N11)*DLOG(P(2,2))

C CALCULATE LIKELIHOODS OF MISSING CENTERS FOR EACH POSSIBLE
C SET OF LINKS
        DO 20 I=1,4
         IF(MISS(I,6).EQ.0)GO TO 20
         LINK1=(I-1)/2+1
         LINK2=MOD(I-1,2)+1
```

```
         PARLIK=0.D0

        DO 21 K=1,64

C       CALCULATE LIKELIHOOD OF POSSIBILITY
           TEMP=P(LINK1,POS1(1,K))
           DO 23 L=1,5
              TEMP=TEMP*P(POS1(L,K),POS1(L+1,K))
23         CONTINUE
           TEMP=TEMP*P(POS1(6,K),LINK2)
           PARLIK=PARLIK+TEMP
21      CONTINUE
        LOGLIK=LOGLIK+DFLOAT(MISS(I,6))*DLOG(PARLIK)
20      CONTINUE


C CALCULATE LIKELIHOODS OF MISSING BEGINNINGS
        DO 30 I=5,6
        LINK2=I-4
        DO 31 J=1,11
        IF(MISS(I,J).EQ.0)GO TO 31
        PARLIK=0.D0
           LIM=2**J
        DO 32 KK=1,LIM

C       CALCULATE LIKELIHOOD OF POSSIBILITY
        K=KK+LIM-2
           TEMP=INIT(POS2(1,K))
        IF(J.EQ.1)GOTO 33
           DO 34 L=1,J-1
              TEMP=TEMP*P(POS2(L,K),POS2(L+1,K))
34         CONTINUE
33         TEMP=TEMP*P(POS2(J,K),LINK2)
           PARLIK=PARLIK+TEMP
32      CONTINUE
        LOGLIK=LOGLIK+DFLOAT(MISS(I,J))*DLOG(PARLIK)
31      CONTINUE
30      CONTINUE

C MULTIPLY BY -1 AND WRITE THE PARAMS AND -1 TIMES THE LOG LIKELIHOOD
        F=-LOGLIK
        WRITE(11,421)X,F
421     FORMAT(1X,3D25.16,1X,D30.20)
        RETURN

C IF THE PARAMS ARE NOT IN THE PARAM SPACE, SET F TO A LARGE NUMBER
666     F=1.D10
        WRITE(11,421)X,F
        RETURN


        ENTRY BOGUS
C  READS IN THE DATA AND GENERATES ALL THE POSSIBLE SEQUENCES OF MISSING
C  DATA

        OPEN(UNIT=10,STATUS='OLD',FILE='P4INP.73')
        READ(10,1)((MISS(I,J),J=1,11),I=1,8)
1       FORMAT(11I4)
```

```
       READ(10,2)N00,N01,N10,N11,N0,N1
2      FORMAT(2I4)

C GENERATE THE 64 POSSIBLE SEQUENCES OF LENGTH 6 CONSISTING
C OF 0'S AND 1'S
       DO 51 K=1,64
          REM=K-1
          POW=5
          DO 52 PL=1,6
             POW2=2**POW
             POS1(PL,K)=REM/POW2+1
             REM=MOD(REM,POW2)
             POW=POW-1
52        CONTINUE
51     CONTINUE


       DO 53 J=1,11
C CALCULATE ALL SEQUENCES OF LENGTH J OF 0'S AND 1'S
       LIM=2**J
          DO 54 K=1,LIM
             REM=K-1
             POW=J-1
             KK=K+LIM-2
             DO 55 PL=1,J
                POW2=2**POW
                POS2(PL,KK)=REM/POW2+1
                REM=MOD(REM,POW2)
                POW=POW-1
55           CONTINUE
54        CONTINUE
53     CONTINUE
       RETURN

       END

C MAIN PROGRAM
       EXTERNAL FUNCT
       DOUBLE PRECISION X(3),H(6),G(3),W(9),F
       OPEN(UNIT=11,STATUS='NEW',CARRIAGECONTROL='LIST',FILE='P4.OUT')

C SET NUMBER OF PARAMS, NUMBER OF SIGNIFICANT FIGURES DESIRED, MAXIMUM
C NUMBER OF FUNCTION EVALUATIONS
       NN=3
       NSIG=8
       WRITE(11,976)NSIG
976    FORMAT(1X,'NSIG=',I2)
       MAXFN=500

C INITIALIZE HESSIAN TO THE IDENTITY AND GIVE INITIAL ESTIMATES
C OF PARAMETERS
       IOPT=0
       X(1)=0.5D0
       X(2)=0.5D0
       X(3)=0.5D0
       WRITE(11,765)X(1),X(2),X(3)
765    FORMAT(/1X,'INITIAL VALUES:'/1X,'P0= ',F25.16,'  P00= '
     1  ,F25.16,1X,'P10= ',F25.16/)
```

```
C READ DATA AND CALCULATE ALL OF THE POSSIBLE SEQUENCES OF MISSING DATA
      CALL BOGUS

C CALL ZXMIN TO MINIMIZE -1 TIMES THE LOG LIKELIHOOD
378   CALL ZXMIN(FUNCT,NN,NSIG,MAXFN,IOPT,X,H,G,F,W,IER)

C WRITE THE RESULTS
      WRITE(11,91)X
91    FORMAT(/1X,'P0=',D25.16,'  P00=',D25.16/' P10=',D25.16/)
      WRITE(11, 92)H
92    FORMAT(1X,'H= ',3D25.16/4X,3D25.16/)
      WRITE(11,93)IER
93    FORMAT(1X,'IER=',I5/)
      WRITE(11,94)F,G
94    FORMAT(1X,'F=',D25.16/1X,'G=',3D25.16/)
      WRITE(11,95)(W(KKK),KKK=1,3)
95    FORMAT(1X,'W= ',3D25.16)
      STOP
      END
```

Program 5

This program calculates the Hessian of the Markov log likelihood.

```
CCCCCCCCCCCCCCCCCCCCCCCCCCCCCCCCCCCCCCCCCCCCCCCCCCCCCCCCCCCCCCCCCCCC
C   THIS PROGRAM EVALUATES THE SECOND DERIVATIVES OF THE LOG LIKELI-  C
C   HOOD FOR THE MARKOV MODEL AT THE MLES.                            C
C                                                                     C
C   IO:                                                               C
C       10 - INPUT                                                    C
C       11 - OUTPUT                                                   C
C                                                                     C
C   NOTATION:                                                         C
C       DLLP0 : FIRST DERIV OF LOG LIKELIHOOD WRT P0                  C
C       DLLP00 : FIRST DERIV OF LOG LIKELIHOOD WRT P00                C
C       DLLP10 : FIRST DERIV OF LOG LIKELIHOOD WRT P10                C
C       D2L0 : SECOND DERIV OF LOG LIKELIHOOD WRT P0                  C
C       D2L0000 : SECOND DERIV OF LOG LIKELIHOOD WRT P0 AND P00       C
C       D2L010 : SECOND DERIV OF LOG LIKELIHOOD WRT P0 AND P10        C
C       D2L00 : SECOND DERIV OF LOG LIKELIHOOD WRT P00                C
C       D2LB : SECOND DERIV OF LOG LIKELIHOOD WRT P00 AND P10         C
C       D2L10 : SECOND DERIVATIVE OF LOG LIKELIHOOD WRT P10           C
C                                                                     C
C                                                                     C
CCCCCCCCCCCCCCCCCCCCCCCCCCCCCCCCCCCCCCCCCCCCCCCCCCCCCCCCCCCCCCCCCCCC


        DOUBLE PRECISION INIT(2),P(2,2),LOGLIK,DLLP0,DLLP00,DLLP10
        DOUBLE PRECISION D2L0,D2L000,D2L010,D2L00,D2LB,D2L10
        DOUBLE PRECISION D2P000,D2P010,D2P00,D2PB,D2P10
        DOUBLE PRECISION T1,T2,T3,T4,T5,T6,T7,T8,T9,T10,T11,T12,T13
        DOUBLE PRECISION DP0,DP00,DP10,PARLIK,PAR2,TEMP
        DOUBLE PRECISION DN0,DN1,DN00,DN01,DN10,DN11,DMISS
        INTEGER POSS(11),POWER,REM,PLACE,MISS(8,11),POW2
        INTEGER M(2,2)
        OPEN(UNIT=10,STATUS='OLD',FILE='P5INP.73')
        OPEN(UNIT=11,STATUS='NEW',CARRIAGECONTROL='LIST',
        FILE='P5.OUT')

C READ INPUT
        READ(10,1)((MISS(I,J),J=1,11),I=1,8)
1       FORMAT(11I4)
        READ(10,2)N00,N01,N10,N11,N0,N1
2       FORMAT(2I4)

C SET PARAMS TO MLES
        P(1,1)=.95923541D+0
        P(2,1)=.86485064D+0
        INIT(1)=.96047966D+0
        P(1,2)=1.D0-P(1,1)
        P(2,2)=1.D0-P(2,1)
        INIT(2)=1.D0-INIT(1)
        WRITE(11,700)P(1,1)
700     FORMAT(1X,' P00 = ',D26.15)
        WRITE(11,701)P(2,1)
701     FORMAT(1X,' P10 = ',D26.15)
```

```
        WRITE(11,702)INIT(1)
702     FORMAT(1X,' P0 = ',D26.15/)

C CHANGE TO DOUBLE PRECISION
        DN0=DFLOAT(N0)
        DN1=DFLOAT(N1)
        DN00=DFLOAT(N00)
        DN01=DFLOAT(N01)
        DN10=DFLOAT(N10)
        DN11=DFLOAT(N11)
C ADD PROBABILITIES OF OBSERVED TRANSITIONS TO LOG LIKELIHOOD
        LOGLIK=DN0*DLOG(INIT(1))+DN1*DLOG(INIT(2))+DN00*DLOG(P(1,1))
     1   +DN01*DLOG(P(1,2))+DN10*DLOG(P(2,1))+DN11*DLOG(P(2,2))

C TAKE DERIVATIVES OF THIS PART OF THE LOG LIKELIHOOD
        DLLP0=DN0/INIT(1)-DN1/INIT(2)
        DLLP00=DN00/P(1,1)-DN01/P(1,2)
        DLLP10=DN10/P(2,1)-DN11/P(2,2)
        D2L0=-1.D0*(DN0/INIT(1)**2+DN1/INIT(2)**2)
        D2L000=0.D0
        D2L010=0.D0
        D2L00=-1.D0*(DN00/P(1,1)**2+DN01/P(1,2)**2)
        D2LB=0.D0
        D2L10=-1.D0*(DN10/P(2,1)**2+DN11/P(2,2)**2)


C CALCULATE LIKELIHOODS OF MISSING CENTERS FOR EACH POSSIBLE
C   SET OF LINKS
        DO 20 I=1,4
        IF(MISS(I,6).EQ.0)GO TO 20
        LINK1=(I-1)/2+1
        LINK2=MOD(I-1,2)+1
        PARLIK=0.D0

C INITIALIZE SOME TEMPORARY VARIABLES
        DP0=0.D0
        DP00=0.D0
        DP10=0.D0
        D2P00=0.D0
        D2PB=0.D0
        D2P10=0.D0


C       GENERATE THE 64 POSSIBILITIES FOR SEQUENCES OF LENGTH 6
C       CONSISTING OF 1'S AND 0'S
        DO 21 K=1,64
           REM=K-1
           POWER=5
           DO 22 PLACE=1,6
              POW2=2**POWER
              POSS(PLACE)=REM/POW2+1
              REM=MOD(REM,POW2)
              POWER=POWER-1
22         CONTINUE

C       CALCULATE LIKELIHOOD OF POSSIBILITY
           TEMP=P(LINK1,POSS(1))
```

```
C       CALCULATE MATRIX OF TRANSITIONS FOR THIS SEQUENCE
           M(1,1)=0
           M(1,2)=0
           M(2,1)=0
           M(2,2)=0
           T4=1.D0
           T5=1.D0
           M(LINK1,POSS(1))=M(LINK1,POSS(1))+1

           DO 23 L=1,5
              TEMP=TEMP*P(POSS(L),POSS(L+1))
              M(POSS(L),POSS(L+1))= M(POSS(L),POSS(L+1))+1
23         CONTINUE

           TEMP=TEMP*P(POSS(6),LINK2)
           M(POSS(6),LINK2)=M(POSS(6),LINK2)+1
           PARLIK=PARLIK+TEMP

C       CALCULATE DERIVATIVES
           T4=T4*(P(2,1)**M(2,1))*(P(2,2)**M(2,2))*
     1    (-1.D0*(P(1,1)**M(1,1))*DFLOAT(M(1,2))*
     1    (P(1,2)**(M(1,2)-1))+DFLOAT(M(1,1))*
     1    (P(1,1)**(M(1,1)-1))*(P(1,2)**M(1,2)))
           T5=T5*(P(1,1)**M(1,1))*(P(1,2)**M(1,2))*
     1    (-1.D0*(P(2,1)**M(2,1))*DFLOAT(M(2,2))*
     1    (P(2,2)**(M(2,2)-1))+DFLOAT(M(2,1))*
     1    (P(2,1)**(M(2,1)-1))*(P(2,2)**M(2,2)))
           DP00=DP00+T4
           DP10=DP10+T5
           T6=(P(2,1)**M(2,1))*(P(2,2)**M(2,2))*
     1      ((P(1,1)**M(1,1))*DFLOAT(M(1,2))*DFLOAT((M(1,2)-1))*
     1      (P(1,2)**(M(1,2)-2))-2.D0*DFLOAT(M(1,1))*DFLOAT(M(1,2))
     1      *(P(1,1)**(M(1,1)-1))*(P(1,2)**(M(1,2)-1))
     1      +DFLOAT(M(1,1))*DFLOAT((M(1,1)-1))*(P(1,1)**(M(1,1)-2))*
     1      (P(1,2)**M(1,2)))
           D2P00=D2P00+T6
           T7=(-1.D0*(P(2,1)**M(2,1))*DFLOAT(M(2,2))*(P(2,2)**
     1      (M(2,2)-1))
     1      +DFLOAT(M(2,1))*(P(2,1)**(M(2,1)-1))*(P(2,2)**M(2,2)))
     1      *(-1.D0*(P(1,1)**M(1,1))*DFLOAT(M(1,2))*(P(1,2)
     1      **(M(1,2)-1))
     1      +DFLOAT(M(1,1))*(P(1,1)**(M(1,1)-1))*(P(1,2)**M(1,2)))
           D2PB=D2PB+T7
           T8=(P(1,1)**M(1,1))*(P(1,2)**M(1,2))*(P(2,1)**M(2,1)
     1      *DFLOAT(M(2,2))*DFLOAT((M(2,2)-1))*(P(2,2)**(M(2,2)-2))
     1      -2.D0*
     1      DFLOAT(M(2,1))*DFLOAT(M(2,2))*(P(2,1)**(M(2,1)-1))
     1      *(P(2,2)**(M(2,2)-1))
     1      +(P(2,2)**M(2,2))*DFLOAT(M(2,1))*DFLOAT(M(2,1)-1)*
     1      (P(2,1)**(M(2,1)-2)))
           D2P10=D2P10+T8

21      CONTINUE

C ADD TO LOG LIKELIHOOD
        DMISS=DFLOAT(MISS(I,6))
        LOGLIK=LOGLIK+DMISS*DLOG(PARLIK)
```

```
C ADD TO DERIVATIVES
        DLLP00=DLLP00+DMISS*(DP00/PARLIK)
        DLLP10=DLLP10+DMISS*(DP10/PARLIK)
        PAR2=PARLIK**2
        D2L00=D2L00+DMISS*(PARLIK*D2P00-DP000**2)/PAR2
        D2LB=D2LB+DMISS*(PARLIK*D2PB-DP00*DP10)/PAR2
        D2L10=D2L10+DMISS*(PARLIK*D2P10-DP10**2)/PAR2

20      CONTINUE

C CALCULATE LIKELIHOODS OF MISSING BEGINNINGS
        DO 30 I=5,6
        LINK2=I-4
        DO 31 J=1,11
        IF(MISS(I,J).EQ.0)GO TO 31

C INITIALIZE TEMPORARY VARIABLES
        PARLIK=0.D0
        DP0=0.D0
        DP00=0.D0
        DP10=0.D0
        D2P000=0.D0
        D2P010=0.D0
        D2P00=0.D0
        D2PB=0.D0
        D2P10=0.D0

C GENERATE POSSIBLE SEQUENCES OF LENGTH J
        LIM=2**J
        DO 32 K=1,LIM
           REM=K-1
           POWER=J-1
           DO 33 PLACE=1,J
              POW2=2**POWER
              POSS(PLACE)=REM/POW2+1
              REM=MOD(REM,POW2)
              POWER=POWER-1
33         CONTINUE

C       CALCULATE LIKELIHOOD OF POSSIBILITY
           TEMP=INIT(POSS(1))

C       CALCULATE MATRIX OF TRANSITIONS FOR THIS SEQUENCE  AND INITIALIZE
C          SOME TEMPORARY VARIABLES
           M(1,1)=0
           M(1,2)=0
           M(2,1)=0
           M(2,2)=0
           T1=-2.D0*DFLOAT(POSS(1))+3.D0
           T2=TEMP
           T3=TEMP
           T9=T1
           T10=T1
           T11=TEMP
           T12=TEMP
           T13=TEMP
```

```
          IF(J.EQ.1)GO TO 39
          DO 34 L=1,J-1
             TEMP=TEMP*P(POSS(L),POSS(L+1))
             M(POSS(L),POSS(L+1))= M(POSS(L),POSS(L+1))+1
34        CONTINUE

39        TEMP=TEMP*P(POSS(J),LINK2)
          M(POSS(J),LINK2)=M(POSS(J),LINK2)+1
          PARLIK=PARLIK+TEMP

C     CALCULATE DERIVATIVES
          T1=T1*(P(1,1)**M(1,1))*(P(1,2)**M(1,2))*
     1    (P(2,1)**M(2,1))*(P(2,2)**M(2,2))
          DP0=DP0+T1
          T2=T2*(P(2,1)**M(2,1))*(P(2,2)**M(2,2))*
     1    (-1.D0*(P(1,1)**M(1,1))*DFLOAT(M(1,2))*
     1    (P(1,2)**(M(1,2)-1))+DFLOAT(M(1,1))*
     1    (P(1,1)**(M(1,1)-1))*P(1,2)**M(1,2))
          DP00=DP00+T2
          T3=T3*(P(1,1)**M(1,1))*(P(1,2)**M(1,2))*
     1    (-1.D0*(P(2,1)**M(2,1))*DFLOAT(M(2,2))*
     1    (P(2,2)**(M(2,2)-1))+DFLOAT(M(2,1))*
     1    (P(2,1)**(M(2,1)-1))*(P(2,2)**M(2,2)))
          DP10=DP10+T3
          T9=(P(2,1)**M(2,1))*(P(2,2)**M(2,2))*T9*(-1.D0*
     1    (P(1,1)**M(1,1))*DFLOAT(M(1,2))*(P(1,2)**(M(1,2)-1))+
     1     DFLOAT(M(1,1))*(P(1,1)**(M(1,1)-1))*(P(1,2)**M(1,2)))
          D2P000=D2P000+T9
          T10=(P(1,1)**M(1,1))*(P(1,2)**M(1,2))*T10*
     1    (-1.D0*P(2,1)**M(2,1)*DFLOAT(M(2,2))*(P(2,2)**(M(2,2)-1))+
     1    M(2,1)*(P(2,1)**(M(2,1)-1))*(P(2,2)**M(2,2)))
          D2P010=D2P010+T10
          T11=T11*(P(2,1)**M(2,1))*P(2,2)**M(2,2)*(DFLOAT(M(1,2))*
     1    DFLOAT(M(1,2)-1)*(P(1,1)**M(1,1))*(P(1,2)**(M(1,2)-2))-
     1    2.D0*DFLOAT(M(1,1))*DFLOAT(M(1,2))*(P(1,1)**(M(1,1)-1))
     1    *(P(1,2)**(M(1,2)-1))
     1    +DFLOAT(M(1,1))*DFLOAT(M(1,1)-1)*(P(1,1)**(M(1,1)-2))*
     1    (P(1,2)**M(1,2)))
          D2P00=D2P00+T11
          T12=T12*(-1.D0*P(1,1)**M(1,1)*DFLOAT(M(1,2))*
     1     (P(1,2)**(M(1,2)-1))
     1    +DFLOAT(M(1,1))*(P(1,1)**(M(1,1)-1))*P(1,2)**M(1,2))*
     1    (-1.D0*P(2,1)**M(2,1)*DFLOAT(M(2,2))*(P(2,2)**(M(2,2)-1))+
     1    DFLOAT(M(2,1))*(P(2,1)**(M(2,1)-1))*P(2,2)**M(2,2))
          D2PB=D2PB+T12
          T13=T13*(P(1,1)**M(1,1))*(P(1,2)**M(1,2))*(DFLOAT(M(2,2))*
     1    DFLOAT(M(2,2)-1)*(P(2,1)**M(2,1))*(P(2,2)**(M(2,2)-2))
     1    -2.D0*DFLOAT(M(2,1))*DFLOAT(M(2,2))*(P(2,2)**(M(2,1)-1))*
     1    (P(2,2)**(M(2,2)-1))+DFLOAT(M(1,1))*DFLOAT(M(1,1)-1)*
     1    (P(1,1)**(M(1,1)-2))*(P(1,2)**M(1,2)))
          D2P10=D2P10+T13

32    CONTINUE

C ADD TO LOGLIK
      DMISS=DFLOAT(MISS(I,J))
      LOGLIK=LOGLIK+DMISS*DLOG(PARLIK)
```

```
C ADD TO DERIVATIVES
        DLLP0=DLLP0+DMISS*(DP0/PARLIK)
        DLLP00=DLLP00+DMISS*(DP00/PARLIK)
        DLLP10=DLLP10+DMISS*(DP10/PARLIK)
        D2L0=D2L0-DMISS*DP0**2
        PAR2=PARLIK**2
        D2L000=D2L000+DMISS*(PARLIK*D2P000
     1   -DP0*DP00)/PAR2
        D2L010=D2L010+DMISS*(PARLIK*D2P010-
     1   DP0*DP10)/PAR2
        D2L00=D2L00+DMISS*(PARLIK*D2P00-
     1   DP00**2)/PAR2
        D2LB=D2LB+DMISS*(PARLIK*D2PB-
     1   DP00*DP10)/PAR2
        D2L10=D2L10+DMISS*(PARLIK*D2P10-
     1   -DP10**2)/PAR2

 31     CONTINUE
 30     CONTINUE


C WRITE RESULTS
        WRITE(11,3)LOGLIK
 3      FORMAT(1X,D25.16/)
        WRITE(11,4)DLLP0,DLLP00,DLLP10
 4      FORMAT(1X,' GRAD = ',3D25.16/)
        WRITE(11,5)D2L0,D2L000,D2L00,D2L010,D2LB,D2L10
 5      FORMAT(1X,' HESS = ',3D25.16/9X,3D25.16)
        STOP
        END
```

REFERENCES

Alexander, C.H. (1981). "Plans to produce National Crime Survey Estimates of 'residences touched by crime'". Unpublished Census Bureau memorandum.

Alexander, C.H., Roebuck, M.J. (1983) "National Crime Survey: Empirical and Model-Based Estimators for the Proportion of Households Victimized in a Year". Unpublished Bureau of the Census memorandum.

Bishop, Y.M.M., Fienberg, S.E., Holland, P.W. (1975). *Discrete Multivariate Analysis: Theory and Practice*. The MIT Press, Cambridge, MA.

Bixby, T.G., Brevoort, S.L., Marks, G.A. (1979). *OHDS - An Introduction to OSIRIS Hierarchical Data Structures*, Ann Arbor, Michigan: Inter-University Consortium for Political and Social Research, University of Michigan.

Cochran, W.G. (1977). *Sampling Techniques*, John Wiley and Sons, New York.

Cox, D.R. and Hinkley, D.V. (1974). *Theoretical Statistics*, Chapman and Hall, London.

Dawid, A.P. (1979). "Some Misleading Arguments Involving Conditional Independence", *Journal of the Royal Statistical Society B*. 41, No.2, pp. 249-252.

Dodge, R.W., Lentzner, H.R. (1980). Patterns of personal series incidents in the National Crime Survey. *Review of Public Data Use*, Vol. 8, pp. 27-41

Dodge, R.W. (1981). The Washington, D.C. recall study, In *The National Crime Survey: Working Papers, Vol.1*. (eds. R.G. Lehnen and W.G. Skogan) Washington, D.C., U.S. Government Printing Office.

Eddy, W.F., Fienberg, S.E., Griffin, D., and Trader, D. (1981a). "NCS Data Files Available at CMU." Working Memorandum NCS-1, Department of Statistics, Carnegie-Mellon University.

Eddy, W.F., Fienberg, S.E., Griffin, D.L. (1981b). "Some First Attempts to Provide Cheery Indicators of Crime Using NCS Victimization Data - Revised Version." Working Memorandum NCS-2R, Department of Statistics, Carnegie-Mellon University.

Eddy, W.F., Fienberg, S.E., Griffin, D.L. (1981c). "Estimating Victimization Prevalence in a Rotating Panel Survey." Bulletin of the International Statistics Institute, 42nd Session.

Eddy, W.F., Fienberg, S.E., Griffin, D.L. (1982). "Longitudinal Models, Missing Data, and the Estimation of Victimization Prevalence." In *Quantitative Criminology; Innovations. and Applications* (J, Hagan, ed.), Sage Research Progress Series in Criminology.

Fay, R. (1982a). "Contingency Table Analysis for Complex Sample Designs: CPLX", Unpublished U.S. Bureau of the Census memorandum.

Fay, R. (1982b). "Contingency Table Analysis for Complex Sample Designs (CPLX), Program Documentation", Unpublished U.S. Bureau of the Census memorandum.

Fienberg, S.E. (1978). "Victimization and the National Crime Survey: Problems of Design and Analysis." *Survey Sampling and Measurement* (N.K. Namboodiri, ed.), New York: Academic Press, pp. 89-106.

Fienberg, S.E. (1980a). "Statistical modelling in the analysis of repeat victimization." *Indicators of Crime and Criminal Justice: Quantitative Studies* (S. Fienberg and A. Reiss, Jr., eds.), Washington, D.C.: U.S. Government Printing Office, pp. 54-58.

Fienberg, S.E. (1980b). "The measurement of crime victimization: prospects for panel analysis of a panel survey." *The Statistician 29*, pp. 313-350

Griffin, D.L. (1981). "Discussion of Several Estimators of the Proportion of Households Victimized in a Year." Working Memorandum NCS-3, Department of Statistics, Carnegie-Mellon University.

Hansen, M.H., Hurwitz, W.N., Madow, W.G. (1953). *Sample Survey Methods and Theory, Vol. I: Methods and Applications*, John Wiley and Sons.

Hansen, M.H., Hurwitz, W.N., Madow, W.G. (1953). *Sample Survey Methods and Theory, Vol. II: Theory*, John Wiley and Sons.

Johnson, N.L. and Kotz, S. (1969). *Discrete Distributions.* Houghton Mifflin Co., Boston.

Kadane, J.B. (1983). "Is Victimization Chronic? A Bayesian Analysis of Multinomial Missing Data." Technical Report # 307, Department of Statistics, Carnegie-Mellon University.

Kish, L. (1965). *Survey Sampling*. John Wiley and Sons, New York.

Lehnen, A.G. and Reiss, R.J.,Jr.(1978). "Response effects in the National Crime Survey". *Victimology*, 3, pp.110-124.

Lehnen, A.G. and Skogan, W.G. (eds.)(1981). *The National Crime Survey: Working Papers, Volume 1: Current and Historical Perspectives.* Washington, D.C.: U.S. Government Printing Office.

Little, R. (1980). "Superpopulation models for non-response II: The non-ignorable case." National Academy of Sciences, Committee on National Statistics Panel on Incomplete Data. To be published.

Penick, B.K. and Owens, M.E.B. (eds.)(1976). *Surveying Crime* (Report of Panel for the Evaluation of Crime Surveys), National Academy of Science, Washington, DC.

Reiss, A.J. Jr. (1979). "Population dynamics and attrition bias in panel surveys." Working Paper, Institution for Social and Policy Studies, Yale University, December.

Reiss, A.J. Jr. (1980). "Victim proneness by type of crime in repeat victimization." *Indicators of Crime and Criminal Justice: Quantitative Studies.* Washington, D.C.: U.S. Government Printing Office, pp. 41-53.

Rubin, D.E. (1976). "Inference and missing data." *Biometrika* 63, pp. 581-592

Schervish, M.J. (1983). "A Beta Binomial Model for Victimization." Technical Report 273, Department of Statistics, Carnegie-Mellon University.

Tallis, G.M. (1962). "The use of a generalized multinomial distribution in the estimation of correlation in discrete data." *Journal of the Royal Statistical Society B.* 24, pp. 530-534.

U.S. Bureau of the Census (1978). *The Current Population Survey: Design and Methodology*, Technical Paper 440. Washington D.C.: U.S. Government Printing Office.

U.S. Department of Commerce, Bureau of the Census (undated). *National Crime Survey, National Sample, Survey Documentation.*

U.S. Department of Justice, Bureau of Justice Statistics (1981a). "The Prevalence of Crime." Washington, D.C.: U.S. Government Printing Office.

U.S. Department of Justice, Bureau of Justice Statistics (1981b). *National Crime Surveys: National Sample, 1973-1979.* Ann Arbor, Michigan: Inter-University Consortium for Political and Social Research, University of Michigan.

U.S. Department of Justice, Bureau of Justice Statistics (1981c). *Sourcebook of Criminal Justice Statistics - 1980.* (M.J. Hindelang, M.R. Gottfredson, and T.J. Flanagan, eds.), Washington, D.C.: U.S. Government Printing Office.

U.S. Department of Justice, Bureau of Justice Statistics (1981d). *Criminal Victimization in the United States, 1979.* Washington, D.C.: U.S. Government Printing Office.

U.S. Department of Justice, Bureau of Justice Statistics(1982). "Households Touched by Crime 1981."*Bureau of Justice Statistics Bulletin.* Washington, D.C.: U.S. Government Printing Office.

U.S. Department of Justice, Law Enforcement Assistance Administration (1978). *National Crime Surveys, National Sample, 1971-1977.* Ann Arbor, Michigan: Inter-University Consortium for Political and Social Research, University of Michigan